THE STUDENT'S GUIDE TO

Food & Drink

JOHN COUSINS & ANDREW DURKAN

GW00500495

Hodder & Stoughton
LONDON SYDNEY AUCKLAND

'Strange to see how a good dinner and feasting
reconciles everybody.'
Samuel Pepys

British Library Cataloguing in Publication Data
Cousins, John
  Students Guide to Food and Drink. – New ed
  I. Title II. Durkan, Andrew
  641

  ISBN 0–340–57532–8

First published in 1990 under the title The Headway Guide to Food and Drink
Re-issued in 1992 as The Students Guide to Food and Drink
© 1990 John Cousins, Andrew Durkan and Conrad Tuor

Typeset by Wearside Tradespools, Fulwell, Sunderland
Printed in Hong Kong for Hodder and Stoughton Educational, a division of Hodder and
Stoughton Ltd, Mill Road, Dunton Green, Sevenoaks, Kent by Colorcraft Ltd

Contents

INTRODUCTION

Over the last 100 or so years Britain has seen some remarkable changes in the type and range of food and drink available in shops and in eating places. There has also been a substantial increase in the number of people eating out and a growing interest in the exploration of good food and drink, both in the home and in restaurants. The British public now eats about 23 million meals away from home each day in a catering industry of some 275,000 establishments. Roughly half of these meals are provided in what are known as the *non-profit sectors* (eg hospitals, welfare services, prisons, the Forces, and industrial feeding), whilst the other half are provided in the *profit* sectors.

This book is dedicated to the promotion of eating and drinking, both for pleasure and for those with a serious gastronomic interest. It is presented in three main sections: Section 1 provides a variety of information on eating and drinking, Section 2 deals with food and Section 3 with wine and other drinks. Overall we have endeavoured to provide a source of useful advice and information mainly from a customer's perspective which will help your exploration of the pleasures of food and drink.

This book also considers the meal as a whole, recognizing that food and drink are not the only factors contributing to one's sense of pleasure or, for that matter, displeasure. Where you eat can be just as important as with whom, as can the particular needs you might have at the time. Some information is also provided for those who may be entertaining at home or booking functions.

Above all eating and drinking should be approached with a certain amount of adventure and fun; perhaps with a sense of humour and the realization that it may not always be perfect, may not meet your particular taste; or may not be exactly what you would normally eat or drink.

In preparing this book we have drawn on a wide variety of experience and literature. In particular we should like to thank Victor Ceserani, MBE, former Head of Department, School of Hotel-keeping and Catering, Ealing College, for encouraging us to prepare this text, and the members of the Food and Beverage Service and Food Studies Divisions at the College for their support.

SECTION

1

On eating and drinking

1 *Choices*

INTRODUCTION

In Britain there are many different kinds of food and drink operations which are designed to meet a wide range of types of demand. It is perhaps important, though, to recognize that it is the *needs* people have at the time, rather than the *type* of people they are, that these different operations are designed for. The same person may be a business customer during the week, but a member of a family at the weekend; he or she may want a quick lunch on one occasion, a snack whilst travelling on another or a meal for the family at a further time. Additionally the same person may be booking a wedding or organizing some other special occasion. Clearly there are numerous reasons for eating out, other examples being: to do something different, to try different foods or for sheer convenience because one is already away from the home – out shopping, at the cinema, a conference or an exhibition.

The reasons for eating out vary and, with this, the types of operation that may be appropriate. Differing establishments offer different services, in both range and price as well as service levels. Also the choice offered may be restricted or wide. Basically there are three types of markets in which operations may be meeting demand. These are:

(a) **captive markets** where the customer has no choice, eg hospital patients
(b) **non-captive markets**, eg those people who have a free choice of establishments
(c) **semi-captive markets** where there is some restriction, eg people travelling by air who have a choice of airline but, once the choice is made, are restricted to the food and drink on offer. This category also applies, for instance, to railways, some inclusive term holidays and people travelling on motorways.

FOOD AND RESTAURANT STYLES

Over the past few years a number of terms have been adopted to signify differing types of food and establishment styles. The *Ackerman Martell Guide* identifies these as follows:

Bistro: normally smaller establishments, checked tablecloths, bentwood chairs, cluttered décor, friendly informal staff. Honest, basic and robust cooking, possible course pâtés, thick soups, casseroles served in large portions.

Brasserie: largish, styled room, often with long bar, normally serving one plate items rather than formal meals (though some offer both). Often possible just to have a drink, or coffee, or just a small amount to eat. Traditional dishes include *charcuterie, moules marinières, steak frites*. Service generally by waiters in long aprons and black waistcoats.

Farmhouse cooking: usually simply cooked with generous portions of basic, home-produced fare using good, local ingredients.

Country house hotel cooking: can vary from establishment to establishment but usually modern English style with some influence from classic or even farmhouse style.

Classic/haute cuisine: the classical style of cooking evolved through many centuries, best chronicled by Escoffier. Greater depth of flavour. Style does not necessarily mean the most expensive ingredients – can include simply poached and boiled dishes such as chicken, tongue and offal. Classical presentation can be served at table or plated.

Cordon bleu: derives from the cookery school of the same name, perhaps normally associated with dinner party or private house cooking.

New/modern English/French: draws from the classical style but with new style saucing and the better aspects of *nouvelle* presentation. Plated in the kitchen, allowing the chef the final responsibility for presentation.

Nouvelle cuisine: at its best, a distinctive style and presentation with a lighter and more innovative approach to some standard dishes. Concentrates on subtle flavours and unusual combinations. Plated, often referred to as 'art on the plate'. Has gained an unfortunate reputation for offering inadequate portions and concentrating more on presentation than content, but at its best it can be exceptionally good.

(*The Ackerman Martell Guide* (London: Alfresco Leisure Publications plc, 1989, p. 14))

FOOD PRODUCTION METHODS

The preparation and cooking of food varies from operation to operation. They may be summarized as eight basic methods but, of course, it is likely that a combination of methods will be found in many operations. The eight basic methods are as follows:

Conventional

A term used to describe production utilizing mainly fresh foods and traditional cooking methods.

Convenience

A method of food production utilizing mainly convenience foods.

Call order

A method of food production where foods are cooked to order either from the customer (as in cafeterias), or from waiting staff. The production area often opens onto the customer area.

Continuous flow

A production line approach where different parts of the production process may be separated. Found mainly in *fast food* establishments.

Commisary or centralized

Production is separated from service by distance, time or both (eg for hospital tray meals or for meals on aircraft). Foods are 'held' and distributed to separate service areas.

Cook-chill

A food production, storage and regeneration method utilizing the principle of low temperature control to preserve the quality of processed foods.

Cook-freeze

A production, storage and regeneration method utilizing the principle of freezing to control and preserve the quality of processed foods. It requires special processes to assist the freezing.

Sous vide

A method of production, storage and regeneration utilizing the principle of sealed vacuum to control and preserve the quality of processed foods.

SERVICE METHODS

Service methods vary in the various establishments that offer food and drink. Firstly, perhaps it is important to distinguish between the 'level' and 'standard' of the service. The **level of service** refers to the *extent* to which the service is more or less personal, as well as to the degree of its sophistication. The **standard of service** refers to the *ability* of the operation to do what it intends to do well.

There are 15 service methods found in the Industry. These can be grouped into five main categories:

- Group A: Table service
- Group B: Assisted service
- Group C: Self-service
- Group D: Single point service
- Group E: Specialized service

These groups have been identified by considering the service from the customer's perspective, ie what happens to the customer. The level of the service reduces from Group A to Group D, with Group E containing specialized forms of service, some of which can be at a variety of levels. This becomes clearer if the service methods are grouped under these headings as follows:

— *Group A: Table service* —

The service of food and drink to customers seated at a laid cover. Within this category there are two main types of service: *waiter* and *bar counter*. However 'waiter service' covers a variety of styles of service.

Waiter service

Food and drink is brought to the customer by food and beverage service staff. The customer is seated at a table. There are six variations:

Silver or English service Food is presented and served, mainly using a spoon and fork, to the customer by waiting staff from a food flat or dish. This method is referred to as *English service* because it was developed in England during the Edwardian era.

Family service Main courses are plated with vegetables and potatoes placed in multi-portion dishes on the tables for customers to help themselves.

Plate or American service Food is served pre-plated to customers.

Butler or French service Food is presented on individual food flats or dishes to customers seated at tables so that they can help themselves.

Russian service The table is laid with a display of food and drink for customers to help themselves. Also sometimes, and confusingly, used to indicate *butler* or *guéridon service*.

Guéridon service Food is served onto a customer's plate at a side table or trolley. Guéridon service may also include carving, cooking and flambage, fish filleting, and the preparation of fresh fruit or salads.

> Note: *It is common nowadays for a particular establishment to combine a number of these variations of waiter service.*

Bar counter

Service to customers seated on stools at bar counters (usually U-shaped). Service is usually plated.

— *Group B: Assisted service* —

This type of service is commonly applied in carvery-type operations where some parts of the meal are served to the customers at the table, while other parts are collected by the customer from a buffet. Also, it is often the standard service used at breakfast in many establishments.

Additionally, this term is used to describe buffets where the customer selects food and drink items from a display and/or trays. The food is usually consumed seated at a table, although the customer may also consume the food while standing or seated in a lounge area.

— *Group C: Self-service* —

Self-service by the customer of food and drink items. The term traditionally applies to cafeterias.

Cafeteria

The customer selects hot and cold food and beverages from counters. There are four variations in counter layouts:

Line counter Customers queue in line formation past a service counter, choosing their menu requirements in stages and loading them onto trays. The counter may include a *carrousel* – a revolving, stacked counter which saves space.

Free flow Selection is from counters but in a food service area where the customer moves at will to random service points. Customers usually exit via a till point.

Echelon A series of counters at angles to the customer flow, thus saving space.

Supermarket Island service points within a free flow area. Customers usually exit via a till point.

— *Group D: Single point service* —

Ordering, service of and payment for food and drink all take place at a single point. Consumption is either on the premises or take-away. This type of service is to be found in a wide variety of operations and takes a number of different forms:

Take-away

The customer orders and is served from a single point at a counter, hatch or snack stand. Consumption is usually off the premises, although some take-away establishments have seating facilities. There are two other variations:

Drive-thru A form of take-away where the customer drives in a vehicle past order, payment and collection points.

Fast food Originally the term *fast food* was used to describe service at a counter or hatch where customers received a meal in exchange for a pre-paid ticket. Nowadays it is commonly used to describe a type of establishment offering a limited range menu, fast service and take-away facility with highly specialized equipment and vast customer throughput. (The term is also used to refer to such things as quick cooking, convenience food production, any take-away, or to any popular catering establishment.)

Vending

The provision of food and drink items by means of automatic retailing. Food may displayed in heated, chilled or ambient cabinets. Sometimes this service is associated with the provision of microwave ovens for reheating food items.

Kiosk

Outstation providing service for peak demand or in a specific location.

Food court

A series of autonomous counters where customers may either order and eat, as in *bar counter* (see page 12), or buy from a number of counters to eat in a separate seating area or take-away.

Bar

A term used to describe the selling point and consumption area in licensed premises.

— *Group E: Specialized service* —

Service where the food and drink are *taken to* the customer; in the previous groups the customer goes to where the food and drink is provided. This group of service methods may also be referred to as *in situ* service.

Tray

A method of service where the whole or part of the meal is served on a tray to customers. It is used in hospitals, on aircraft, on some trains and also in off-premises catering.

Trolley

The service of food and drink from a trolley, away from dining areas. This form of service may be used in offices and is also found on some trains.

Home delivery

Food is delivered to the customer either at home or work. The term also encompasses such services as 'Meals on Wheels'.

Lounge

The service of a variety of food and drinks in the lounge area of an establishment.

Room

The service of a variety of food and drinks in guest apartments or meeting rooms.

Drive-in

Customers park their motor vehicles and are served at the vehicles.

Banquet/function catering or service is used to describe catering for specified numbers of people at specific times, in a variety of dining layouts. Service methods vary. As such this is not a separate form of service; the term referring to the organisation of the service.

LIQUOR LICENSING

In order to sell intoxicating liquor in Britain, licences are required. These licences govern the *type of liquor* which can be sold, the *extent of the market* which can be served and the *times of opening* (or *permitted hours*). There are also regulations on restrictions to young persons and on measures. From the customer's point of view it is important to remember that these restrictions are made by the government not by the licensee, and that penalties for infringement are applied, not only to the licensee, but also to the customer.

— *Types of licence* —

The various types of licence available in Britain are described below (but see the note regarding Scotland).

Full On-Licence

This allows the licensee to sell all types of intoxicating liquor for consumption on and off the premises. However there are a few examples of On-Licences where the type of alcohol is restricted, eg beer only or beer and wine only.

Restricted On-Licences

Restaurant Licence This applies to the sale of alcoholic liquor to persons taking main meals only.

Residential Licence This applies to the sale of alcoholic liquor to persons residing on the premises or to their private friends who are being genuinely entertained by the guests at the guests' expense.

Combined Licence This is a combined Restaurant and Residential Licence.

Club Licences

Licensed Club Normally this is a licence to run a club, which is operated by individuals or a limited company, as a commercial enterprise. The sale of alcoholic liquor is to members only.

Members' Club A licence to run a club, normally by a committee of members, as a non-profit making organization. The members own the stock of liquor and sale is to members only.

Off-Licence

A licence authorising the sale of intoxicating liquor for consumption off the premises only.

Occasional Licence

This is granted to holders of On-, Restaurant or Combined Licences enabling them to sell alcoholic liquor at another place for a specified time, eg a licensee may be able to set up a bar for a local village hall function.

Occasional Permission

This is similar to an Occasional Licence but may be applied for by non-licence holders, eg a charity may apply for Occasional Permission in order to sell alcoholic drink at a specific fund-raising event.

Music and Dancing Licences

These licences are not liquor licences but are required for public music and dancing. The licences are granted by local councils and the law varies from place to place. Licences are not required where radio, television, and recorded music are used or where there are no more than two live performers, although if dancing takes place a licence is required.

— *Permitted hours* —

Currently *permitted hours* are as follows in England and Wales:

Weekdays	11 am to 11pm
	8 am to 11 pm at off-licences
Sundays *Good Friday* *Christmas Day*	12 noon to 3 pm and 7 pm to 10.30 pm at on- and off-licences

Within these permitted hours the licensee can choose when and for how long to close the premises.

Exceptions to permitted hours

The following exceptions apply to permitted hours:

(a) the first 20 minutes after the end of permitted hours is for consumption only
(b) the first 30 minutes after the end of permitted hours for those taking table meals is again for consumption only
(c) residents and their guests may be (but do not have to be) served at any time as long as only the resident makes the purchase.

Note: *Permitted hours in Scotland are similar to those above.*

Extensions to permitted hours

Special Order of Exemption This is available for specific occasions, eg a wedding, dinner dance or carnival.

General Order of Exemption This applies to an area where a particular trade or calling is going on, eg market day or food markets which are operating early in the morning.

Supper Hour Certificate This allows for an additional hour at the end of permitted hours for licensed restaurants.

Extended Hours Certificate This is an extension for establishments which already hold a Supper Hours Certificate and provide some form of entertainment. The extension is until 1 am.

Special Hours Certificate This allows for extensions of permitted hours to premises which are licensed, hold a Music and Dancing Licence and provide substantial refreshment. The extension can be until 3 am in the West End of London and until 2 am elsewhere.

— *Other legal requirements* —

As well as needing a licence, those wishing to sell intoxicating liquor in Britain are bound by other legal requirements relating specifically to young people and to the measure of liquid sold.

Young persons

It is an offence for persons under 18 to be served in a licensed bar. It is also an offence to allow persons under 18 to consume alcoholic beverages in a bar. Similarly, it is an offence for the person under 18 to attempt to purchase or to purchase or consume alcoholic beverages in a bar. The position regarding young persons may be summarized as follows:

Age	Purchase in a bar	Consume in a bar	Enter a bar	Purchase in a restaurant	Consume in a restaurant
Under 14	No	No	No	No	Yes[1]
Under 16	No	No	Yes	No	Yes[1]
Under 18	No	No	Yes	Yes[2]	Yes

[1] As long as the alcoholic beverage is bought by a person over 18
[2] Beer, cider and perry only

Note: *Tobacco should not be sold to persons under 16.*

Weights and measures

Beer and cider Unless sold in a sealed container, beer and cider may only be sold in measures of ⅓ or ½ pint or multiples of ½ pint. This does not apply to mixtures of two or more liquids, eg a shandy.

Spirits For whisky, gin, vodka and rum where these are sold by the measure, they must be sold in measures of ¼, ⅕ or ⅙ of a gill or in multiples thereof (a gill is ¼ pint). A notice must be displayed in the establishment indicating the measure that is being used. This restriction does not apply to mixtures of three or more liquids, eg for cocktails.

Under an EEC directive, Britain will be introducing metric measures likely to be based on 25 millilitres. This legislation will be phased in so as to allow for a change over a period of time.

Wines Wines sold open in carafes must be sold only in measures of 10 or 20 fluid ounces, 25, 50 or 75 centilitres or 1 litre. For wine by the glass, 11 measures are identified. These are 100, 125, 150, 175, 200 and

250 millilitres and 4, 5, 6, 6 ⅔ and 8 fluid ounces. However, it is still legal to serve wine by the glass in any measure if no specific measure is advertised. The code of practice covering wine measures is likely to become law during 1990. However, many caterers have already adopted the 125 millilitre measure since there are exactly six of these in a 75 centilitre wine bottle.

OTHER DIFFERENCES BETWEEN ESTABLISHMENTS

As well as differing service methods, production methods and food and restaurant styles, there are other differences between establishments. These differences include factors such as the space provided, the capacity of the operation, the price, the range of choice, the other types of customers, the décor, lighting and noise levels, the speed of service, the availability of special choices, the flexibility of service offered and the methods available for making payment. All of these factors, not just the food and drink on offer, contribute to the enjoyment of a meal. Additionally there are other factors such as the location of the establishment that can affect the enjoyment of a meal. Difficulties in getting to the establishment or, conversely, its convenience make for a difference in the attitude that you as customer might have towards that establishment.

One of the most significant differences between establishments is the variation in attitude held by the staff. Service staff have a difficult job; it is complex and highly demanding. There is an interesting paradox, though, in that if the job is done well it is not noticed. Without a doubt, scruffy and unhelpful staff can easily spoil a meal. However, at the same time, well turned out, knowledgeable and helpful staff can make up for deficiencies elsewhere in the operation. This is true whatever the type of operation. Less money does not mean that the staff should be less helpful or well turned out.

THE MEAL

Throughout this section differences between types of operations have been identified. It is important to recognize that meals are not just food and drink. In going out to eat a number of **needs** are being satisfied. These might be:

(a) **physiological** needs, eg to satisfy hunger and thirst, or to satisfy the need for special foods
(b) **economic** needs, eg staying within a certain budget, wanting good value, a convenient location or fast service
(c) **social** needs, eg being out with friends, business colleagues or attending special functions such as weddings
(d) **psychological** needs, eg responding to advertising, wanting to try something new, or simply fulfilling life style needs
(e) **convenience** needs, eg it may not be possible to return home or the desire may be there for someone else to prepare, serve and wash up.

You may be wanting to satisfy all or some of these needs and it is important to recognize that it is the *reason behind* wanting or having to eat out, rather than the food and drink by themselves, that will play an important part in determining the resulting satisfaction or dissatisfaction with the experience. It is quite possible that the motivation to eat out is not to satisfy basic physiological needs at all.

— *Meal experience factors* —

Many different aspects of the meal experience affect one's overall attitude at the end of the meal. You might be satisfied or dissatisfied with the service, the food or the seating arrangements. These aspects are the responsibility of the operation. However you might be satisfied or dissatisfied with the other customers, transport or the weather. These matters are outside the control of the establishment but also have an effect on your enjoyment.

Various research has been carried out over the years in order to identify the factors which contribute to the meal experience as a whole and therefore affect the satisfaction or dissatisfaction that may be gained. These may be summarized as follows:

Food and drink on offer

The range of foods, choice, availability, flexibility for special orders and the quality of the food and drink.

Level of service

Depending on the needs people have at the time, the level of service sought should be appropriate to those needs. For example a romantic night out may call for a quiet table in a top end restaurant, whereas a group of young friends might be seeking more informal service. This factor also takes into account services such as booking and account facilities, acceptance of credit cards and also the reliability of the operation's product.

Level of cleanliness and hygiene

This relates to the premises, equipment and staff. Over the last few years this factor has increased in importance in the customers' minds. The recent media focus on food production and the risks involved in buying food have heightened awareness of health and hygiene aspects.

Perceived value for money and price

Customers have perceptions of the amount they are prepared to spend and relate these to differing types of establishments and operations. However many people will spend more if the value gained is perceived as greater than that obtained by spending slightly less. (Also see the notes on price, cost, value and worth below.)

Atmosphere of the establishment

This is difficult to quantify as it is an intangible concept. However it is composed of a number of factors such as: the décor, lighting, heating, furnishings, acoustics and noise levels, the other customers and the attitude of the staff.

MAKING CHOICES
— *Price, cost, worth and value* —

These four words are often used in relation to hotel and catering operations. Each has a specific meaning and it is useful to take some time to consider them.

- **Price** refers to the amount of money that is required in exchange for the meal
- **Cost** refers to matters besides money, eg the cost of not going somewhere else, the cost of transport, the potential cost in embarrassment within certain types of operations, the personal cost

of feeling you are in the wrong place and also the personal cost of having to complain
- **Worth** is the impression that we as customers get of the quality and/or quantity of the meal in relation to the price but irrespective of the cost associated with it
- **Value** refers to the balance between worth and cost. Good value is where the perceived worth is greater than the costs involved and poor value is where the cost is greater than the perceived worth

In Britain *price* is not the main factor in determining choice. People with the money to pay may avoid certain operations because the *costs* are too great. Such costs could be as simple as the fear of embarrassment. Also, some people fear new experiences and will wait for others to visit establishments before venturing themselves. *Value* is very much a personal issue. Again this comes back to the needs people may be seeking to satisfy. To some, value is dependent on having seclusion and they are prepared to pay for it. To others it refers to the quantity and quality of food. To some value is an inclusive meal package and to others it is the ability to choose separately from a list of different items.

People often ask which is the best restaurant or place to eat. This really is an impossible question to answer. People are different; their experiences, their needs at the time and their expectations vary. The best place is probably the one that satisfies one's needs at the time and, therefore, the measure of this is how well the operation meets these needs. To some people a place where you are known is important and to others the need for exploration or something new is more important. In short this is a personal issue. Some help in making choices is provided by restaurant reviews and the various guides that are available. Always remember, however, that these reflect the personal opinions of the reviewers whose needs may be different from your own.

— *Guidebooks available* —

Guides to eating establishments fall into three basic categories:

(a) Guides which provide factual information and classify establishments with reference to a set of criteria, eg how many bathrooms, when the restaurant is open etc. These include:

- The *Automobile Association Members' Handbook*, published annually by the AA and available to its members
- The *Royal Automobile Club Guide*, published annually by the RAC and available to its members
- *Hotels and Restaurants in Britain* (1989), available from AA Publications, Dunhams Lane, Letchworth, Herts SG6 1LF
- *Tourist Board Guides*, published annually and available from Tourist Board offices

All of these organizations also award stars or crowns. Generally the more stars/crowns awarded, the more facilities there are. However, it should be remembered that many establishments are not included in these guides; lack of recognition does not mean that they are less good.

(b) Guides which give some evaluative measure. These include:

- *The Egon Ronay Hotel and Restaurant Guide*, Alfresco Leisure Publications plc, 35 Tadema Road, London SW10 0P2
- *The Ackerman Martell Guide* (1989), published by Alfresco Leisure Publications plc.
- *The Good Food Guide*, published annually by the Consumers' Association, Castlemead, Gascoyne Way, Hertford SG14 1LH
- *The Michelin Guide*, published annually by Michelin et Cie, Proprietaires-Editeurs, France
- A series of *Where to Eat In* publications, eg 'London', are available from Tourist Board offices.

(c) Guides which are linked to some marketing initiative (although the rules governing entries in these guides ensure some measure of consistency among those included). These include:

- *Prestige Hotels*, available from Prestige Hotels, 353 Strand, London WC2R 0HS
- *Relais and Château*, available from the French Government Tourist Office, Piccadilly, London
- *Les Routiers*, published annually by Ebury Press, Brookmount House, 62–65 Chandos Place, Covent Garden, London WC2N 4NW
- *Derek Johansen's Guide*, published annually by Johansens Ltd, Hobsons Publishing, Bateman Street, Cambridge CB2 1LZ.

Guides and recommendations may be of use in choosing where to eat, but there is nothing like *personal experience*. Today's diner is confronted with a wealth of choice; increased experience will lead to greater knowledge and to the ability to read the signs of a good or poor establishment. The latter will always be a personal distinction and depend as much on one's own experience as on the needs one has at the time.

2 *Etiquette in eating out*

MAKING BOOKINGS

Many establishments require that a booking be made prior to the time of the meal. Customers should remember that these bookings are a form of *contract* and are therefore enforceable. Just because a restaurant is open, this does not mean that they are under any obligation to serve you (unless you are a bona fide traveller and have booked into an hotel). If you contact an establishment and request a booking for a certain number of people at a certain time, you have made an offer which the restaurant might accept or offer an alternative. If you fail to turn up on time, then the table need not be held and the restaurant may seek compensation through litigation. Similarly, if you have booked for four people and only three arrive, the restaurant might seek compensation for a loss of revenue from the fourth person.

Of course things can go wrong and either party may be unable, through unforeseen circumstances, to meet the terms of the contract. In this instance there is no obligation for the contract to continue.

If you are considering going out as part of a large party, it is worth asking the restaurant for a restricted menu or even a set one. There are few restaurants which can cope with a full range of choice for a large party and many will welcome this suggestion, others may make it themselves. Another approach is to collect the menu in advance and ask members of the party to pre-select their order. This can then be telephoned through to the restaurant in advance. Again, many restaurants welcome this approach.

SERVICE STAFF

It has often been said that restaurants are like theatres – but have you ever come across a theatre where the same cast give different performances, not necessarily at the same time, to different audiences?

The skilled service staff should be able to do this, and read the type of approach required by the customer. For example, business people generally require the maximum of attention with the minimum of interruption. A couple who have gone out for a celebration, on the other hand, might like a quiet corner. Alternatively an office party might want some greater involvement from the service staff. The mark of a good establishment is where this can happen.

Service staff are professional people doing a professional job. They are involved in a highly complex operation and should perhaps be treated with as much respect as they are taught to show to the customer. There are, of course, exceptions on both sides, but there is nothing clever in being rude to a server. If you have a complaint, make it. Smart comments are not that smart when you realize that the server has probably heard them all before.

THE HOST AND THE GUEST

If you are being taken out for a meal then it is good form to order all menu and drink items through the host and not to seek service directly from the staff. This applies to parties of two as well as to larger parties (such as at formal functions). The privilege of paying is also the privilege of offering food and drink, as if the host were at home.

In addition to these issues are the various codes of social behaviour. These are detailed in books such as:

- *Debrett's Etiquette and Modern Manners* (1981) published by Debrett's Peerage Ltd.
- *The Complete Book of Etiquette* written by Mary and John Bouldon and published by W. Foulsham and Co., Ltd.
- *Enquire Within Upon Everything* (1987) written by Moyra Bremner and published by Century.

3 Healthy eating and drinking

Nowadays there is a greater awareness of the need to consider health when eating and drinking. More choice and availability whilst being the mark of a developed society leads to a necessity for greater care when selecting food and drink.

FOOD

Much has been published recently about food and its link with health. Moves towards better balanced diets containing fewer animal fats and less salt and sugar and more roughage are having a positive effect on health. However there is a need for some sense here. *Balance* in the diet is more important than slavish adherence to healthy eating rules. Foods such as butter and cream are not *bad* foods; it is *excess* which is the problem, not the food itself.

ALCOHOL

Drinking can also have its problems. The great majority of the population who drink alcohol do so for many reasons: to quench a thirst, as a relaxant or simply because they like the stuff. But there's the rub: some people like it too well. It is reminiscent of the man who thought he had an alcohol problem until he found another bottle!

A *small amount* of alcohol does no harm and can even be beneficial. However, the more you drink and the more frequently you drink, the greater the health risks. Alcohol depresses the brain and nerve function thereby affecting one's judgement, self-control and skills.

Most of the alcohol you drink passes into the bloodstream from where it is rapidly absorbed. (This absorption may be slowed down somewhat if drink is accompanied by food.) Almost all the alcohol must then be burnt up by the liver, with the remainder being disposed of in urine or perspiration. Someone once said that the liver is like a car with just one gear. It can only burn up one unit of alcohol in an hour so, if it has to deal with too much alcohol over the years, it will inevitably suffer damage.

So what are the *sensible* limits we can go to if we wish to avoid damaging our health? Of course, if you stop drinking alcohol you cut out any risk. However, medical expert opinion from the Royal College of Physicians sets the limit at *21 units* a week (spread throughout the week) for men and *14 units* a week (spread throughout the week) for women.

1 unit = ½ pint of ordinary beer or lager
or a glass of sherry (⅓ gill)
or a measure of vermouth or other aperitif (⅓ gill)
or one measure of spirits (⅙ gill)

Note: *In Scotland an ordinary spirit measure is ⅕ gill or 1¼ units. In Ireland it is ¼ of a gill or 1½ units. Extra strength lagers have twice and sometimes three times the strength of ordinary beers. Remember also that many low calorie drinks contain more alcohol than their ordinary equivalents.*

From these guidelines we can see that if a man drinks 36 units or more of alcohol throughout the week (eg 18 pints of beer or 1⅑ bottles of whisky) then he is likely to be damaging his health. Similarly if a woman goes beyond 22 units (say 1½ bottles of sherry) throughout the week, then a risk to her health is likely. Be very careful drinking at home or in somebody else's home. A generous hand can be too generous for your own good, so look carefully at the drink level in your glass.

— *Hangovers* —

George Bernard Shaw once said 'eat at leisure, drink by measure'. This is good advice. However, if you do go too far then remember every pleasure has a price; over indulgence in alcohol leads to hangovers. A hangover is a headache resulting from (a) dehydration of the body caused by the alcohol, (b) the presence of congeners (additives etc) in the drink, and (c) lack of real sleep. This is why we suffer after a big drinking binge.

A hangover is an unpleasant experience and people are constantly searching for effective cures. However, like in so many things, *prevention* is always better than cure:

- It is always advisable to drink less and to eat well before an evening of drinking. Food lines the stomach and acts as a buffer against the war that alcohol wages
- Avoid mixing drinks if possible
- Be wary of concoctions such as laced drinks, dubious punches, wine cups and foul, cheap wine
- Drink lots of water before you go and after you come from a heavy drinking party.

However, if you have failed to heed this advice and you feel absolutely rotten in the morning try one of the following remedies.

1 *Prairie Oyster*: into a glass put 1 egg yolk, 1 measure of brandy, 1 teaspoon wine vinegar, 1 teaspoon Worcester sauce and a little cayenne pepper. Drink in a gulp without breaking the yolk
2 If you do not have all the above ingredients at home, take a brisk walk to the chemist, inhaling deeply on your way, and ask for a dose of kaolin and morphine. This is a rather drastic solution but it works
3 There are many proprietary brand bitters which will also do the trick or at least shock the system in the attempt. Underberg and Fernet Branca are particular favourites (see also pages 216–17)
4 The 'hair of the dog' – well why not, if you believe it will do you good!

4 Wine and food harmony

Nowadays the old maxim of red wine with meat and game, white with fish and poultry has given way to a much more relaxed attitude. People have broken away from the very rigid approach to the marriage of food and wine propounded by the wine pundits of yore to those who would listen. They are now much more inclined to drink what they like, when they like, and are much more open and honest about their wine preferences. If they do not care for red or dry white wines, they are not in the least bit intimidated by the raised eyebrows of the haughty wine waiter as they order the luscious Château Rieussec with their steak – and good for them too!

Hosting a table either at home or in a restaurant presents a different situation and here you can hope that your guests have more conventional tastes. However, it is a well-known fact that our choice of wine and other drinks is influenced by a variety of different factors: the occasion, the time of day and year, the weather, one's mood, the location, the theme of the event, the atmosphere, one's previous experience of the party, the food, the price, marketing and the media, the selling skills of the restaurant personnel and, of course, one's personal ego and preference.

GENERAL GUIDELINES

When selecting a wine to accompany a meal it is important to understand the compatability of *flavours* and *textures*. This can only be gleaned through trial and error. However, there are certain guidelines that are worth observing:

(a) Dry wines should be served before sweeter ones
(b) White wines should come before red wines
(c) Lighter wines should be served before heavier ones
(d) Good wines should appear before great ones

(e) The main course wine should always be finished before the sweet to prevent a clash of two different tastes, which will destroy the pleasant memory of the wine with the main course. This is one of the reasons why the French eat cheese before the sweet; the custom ensures that the main course wine is finished before the sweet course is served

(f) Wine should not really be drunk with certain foods like chocolate, eggs, salads with vinegar dressings, mint sauce and very hot and spicy foods such as curries. If an accompaniment is desired with these foods, then something inexpensive should be chosen since the food will flatten and dull the taste of the wine, if it does not completely overpower it.

However, when contemplating possible food and wine partnerships, remember that no guidelines exist to which there are not exceptions. For example, although fish is usually served with white wine, some dishes, such as heavily sauced salmon, red mullet or a fish such as lamprey (which is traditionally cooked in red wine), can be successfully accompanied by a slightly chilled red Saint Emilion, Pomerol or Mercury. The combinations that prove most successful are those that please the individual.

SUGGESTED COMBINATIONS

Clearly, the matching of food and wine is highly subjective. However, the overall intention should be to provide food and wine which harmonise well together, each enhancing the other's performance. Below are some suggestions:

— *Aperitifs* —

The purpose of an aperitif is to sharpen the palate and to start the gastronomic juices flowing in anticipation of the meal to come. Beware of excessive amounts of strong drinks such as spirits and cocktails. However enjoyable they may be at the time, they will deaden the palate for the wine and food that follow.

The name *aperitif* comes from the Latin *aperitivus* – to open out (in this case it is the gastric juices which are 'opened out' to give an appetite for the meal to come).

- Champagne, sherry, dry white wine, Madeira and vermouths are particularly good aperitifs, but they must not be drunk to excess, otherwise the food takes second place and is not likely to be appreciated
- Gin with a variety of partners seems to be the most popular aperitif nowadays: gin and tonic, gin and It (Italian vermouth), gin and French (French vermouth) and, of course, the Martini cocktail (gin and dry vermouth) please most people. For those who dislike gin, vodka is the natural substitute
- Schnapps and pastis are also popular, as are the various bitters
- Whisky and soda, Kir, Buck's Fizz and sundry cocktails all help whet the appetite and get people in the mood for enjoying the meal
- A more simple approach is to serve the wine that will accompany the first course

— *Hors-d'oeuvre* —

- Fino or Manzanilla sherry
- Sancerre or Gewürztraminer

— *Soups* —

- These do not really require a liquid accompaniment but sherry or dry port or Madeira could be tried
- Consommés, turtle soup and lobster or crab bisque can be uplifted by adding a glass of heated sherry or Madeira before serving

— *Foie gras* —

- Beaujolais or a light, young, red wine
- Some people like sweet wines

— *Cheese omelettes and quiches* —

- Ideally, no wine should be served
- An Alsatian Riesling or Sylvaner is probably the most suitable if wine is required

— *Farinaceous dishes* —

- Nothing is better than Italian red wines such as Valpolicella, Chianti, Barolo, Santa Maddalena, Lago di Caldaro

— *Fish* —

- *Oysters and shellfish*: dry white wines, Champagne, Chablis, Muscadet, Soave and Frascati
- *Smoked fish*: white Rioja, Hock, white Graves, Verdicciho
- *Fish dishes with sauces*: these require fuller white wines such as Vouvray, Montrachet or Yugoslav Riesling
- *Shallow fried, poached or grilled fish*: Vinho Verde, Moselle, Californian Chardonnay, Australian Sémillon or Chardonnay

— *White meats* —

The type of wine to serve is dependent on whether the white meat (chicken, turkey, rabbit, veal and pork) is served hot or cold:

- *Served hot with a sauce or savoury stuffing*: either
  a a rosé such as Anjou or
  b light reds like Beaujolais, New Zealand Pinot Noir, Californian Zinfandel, Saint Julien, Bourg and Burgundy, e.g. Passe-tout-grains, and Corbières
- *Served cold*: fuller white wines such as Hocks, Gran Viña Sol, Sancerre and the rosés of Provence and Tavel

— *Other meats* —

- *Duck and Goose*: big red wines that will cut through the fat, eg Châteauneuf-du-Pape, Hermitage, Barolo and the Australian Cabernet Shiraz
- *Roast and grilled lamb*: Médoc, Saint Emilion, Pomerol and any of the Cabernet Sauvignons
- *Roast beef and grilled steaks*: big red Burgundies, Rioja, Barolo, Dão and wines made from the *Pinot Noir* grape
- *Meat stews*: lighter reds, eg Zinfandel, Côtes du Rhône, Clos du Bois, Bull's Blood
- *Hare, venison and game*: reds with distinctive flavour, eg Côte Rôtie, Bourgeuil, Rioja, Chianti, Australian Shiraz, Californian Cabernet, Chilian Cabernet Sauvignon and fine red Burgundies
- *Oriental foods, Peking duck, mild curry, tandori chicken, shish kebab*: Gewürztraminer, Lutomer Riesling, Vinho Verde, Mateus Rosé or Anjou Rosé

— Cheese —

The wine from the main course is often followed through to the cheese course but, if not, almost any wine will do as cheese and wine go together like bread and butter. Having said that, you should still consider the type of cheese being served:

- The *light, cream cheeses* go well with full bodied whites, rosés and light reds
- The *strong, pungent* (even smelly) and *blue veined varieties* cry out for big reds like Bordeaux and Burgundy, or tawny, vintage or vintage-style ports and even luscious sweet whites

— Sweets and puddings —

Most sweets and puddings are only barely comfortable with wines, perhaps because two sweet tastes in the mouth are almost too much of a good thing. However, certain wines can be recommended:

- Champagne works well with sweets and puddings
- The luscious Muscats (de Beaumes-de-Venise, de Sétubal, de Frontignan, Samos), Sainte-Croix-du-Mont, Sauternes, Banyuls, Monbazillac, Tokay and wines made from late gathered individual grapes in Germany all make a brave effort to satisfy. Despite this, it is perhaps preferable to save the wine until after the sweet course when it can be appreciated to the full

— Dessert (fresh fruit and nuts) —

- Sweet fortified wines, sherry, port, Madeira, Malaga, Marsala, Commanderia, Yalumba Galway Pipe and Seppelt's Para

— Coffee —

- Cognac and other brandies such as Armagnac, Asbach, Marc, Metaxa, Grappa, Oude Meester, Fundador, Peristiani VO31
- Good aged malt whiskies
- Calvados, sundry liqueurs and ports

5 Special occasions and functions

Occasionally you may have to arrange a special function; perhaps a wedding, a business dinner or a party of some kind. Below, a number of issues are identified that will need to be considered when making these arrangements.

BUDGET

The first thing to do is to establish a *budget* for the event. This often involves obtaining information from establishments in order to ascertain the likely costs of a function. Many hotels, banqueting houses and some restaurants frequently provide *function packages*. These packages contain information on facilities, room charges, the type of food and drink available together with costs, and other services that can be provided such as live bands, other entertainment, printing services and car hire. Such packages have the advantage of offering a standard range of choices at differing price levels. However, since they may vary with the type of function, it is worthwhile specifying the sort of function you are organizing (eg a wedding) when you make the request.

Although it may be an advantage to take one of these standard packages (the establishment is likely to do it well), it may not cater for all your requirements. Of course, most establishments allow for some flexibility, but this may increase the cost as the function is now being custom designed. In general, however, it is a good idea to obtain a number of function packages since these can give a fairly accurate guide to what is available and the likely costs.

WHERE TO GO

Once the packages have been obtained, or when you have decided where you would like to go, it is always worthwhile visiting the establishment for preliminary discussions. The sooner that this can be done the better. It is not uncommon for establishments to be booked a year ahead, especially for weekends. It is also worthwhile having a meal in the establishment and seeking comments from others whom you know have used the place for function purposes. Bear in mind, however, that a function meal may not always be as good as a small meal for two.

Furthermore, try to cultivate a good relationship with the person at the establishment who will be looking after the function. Always try to deal with one specific named individual. This will help when last minute changes need to be made and will also give you more confidence about the success of the event.

When choosing the venue, do consider the *location, access, ease of parking* or *access to public transport*. If the place is not that well known, it is worthwhile considering sending a map of the location to your intended guests.

FOOD AND DRINK

Most establishments nowadays should be able to cater for special needs such as vegetarian or low fat diets, offering suitable alternatives within the menus available. Few establishments will allow you to bring your own drink and, if they do, will charge corkage which may make it uneconomical.

Take time to agree the menu and the drink. Many establishments offer a complete package at a per person charge. These have the advantage of being priced in such a way that you know exactly what the total cost will be at the end. Other packages are based on a basic price with extras added for any additions that you require.

As well as agreeing the menu, most operators will have some form of function checklist to discuss with you. Again this will give you some confidence. However, do not be pressurized; have what you want but, at the same time, take the advice of the people dealing with you. They have experience of organizing functions and will generally try to ensure that yours goes well.

ADMINISTRATION

Once the establishment has been booked and the food and wine decided upon, there are other, more administrative tasks to be done such as preparing invitations, menus, table lists and plans and name cards. Few places will organize these for you but will recommend local people who can. Additional items such as flowers, entertainment, music, changing rooms, the availability of a master of ceremonies or toast master will usually be organized by the establishment at an additional cost. In some places additional liquor licensing arrangements will be necessary for the event to take place. This will be handled by the establishment, but do check when alcoholic beverages will be available.

It is a good idea to summarize the plan of the event in a chronological order starting from the initial meetings through to invitations, letting the establishment know of the final numbers, arrival times, eating times, and departure times. Do check what time the establishment expects you and your guests to leave as they may have other functions on and it is better to arrange the departure time in advance rather than risk overlapping with another function. It is also worthwhile trying to think of what might go wrong and to make plans for such eventualities. Many establishments will ask for some form of deposit and you should allow for this in your cash availability.

Formal functions are the subject of many traditions, eg the Loyal Toast, orders of precedence, addressing dignitaries, the top table layup for a wedding. Information on these can usually be supplied by the establishment. However, further information can be found in the books listed on page 26.

6 *When things go wrong*

Sometimes things can go wrong. This happens for a variety of reasons, but it is important to remember, from a customer's point of view, that very little, if any, of this is intentional. It just so happens that when one thing goes wrong, it is frequently followed by another and perhaps another, usually to the same person or party of guests. However, when things do go wrong, a complaint is necessary. In Britain people seem too reticent to make complaints so that, by the time a complaint is made, annoyance has tended to set in and the whole business can become unpleasant.

Perhaps one of the problems is that, in Britain, many subscribe to the view that the customer should play a passive role in eating out, especially in establishments offering high levels of service. Sadly there are some establishments where the customer's role is merely to provide money for the establishment to do what it wants. A sort of expensive silence develops where the customers behave as if they are in some revered presence. Such establishments may appeal to some people, but always ensure that they are correct for *you*.

In the same way, ask yourself if you enjoy going to a place where you are only treated as important if you are 'somebody' – if you are not 'somebody', then you are only there as a spectator for those on whom the attention is lavished. If you like such a place, fine, but always be clear that this is *why* you are going to it. How many times have you sat in a restaurant and quietly accepted things, good or bad?

Is there really a right way to do things? Yes, there are a few conventions, but after that who is to say that something is right? But it may be important to consider why someone is saying it is right. Should customers be more clear about what they expect? Should the operation seek out these expectations and attempt to meet them? Some operations do and do it successfully. In an increasingly growing market that is estimated to become highly competitive within the next five years, the only operations that will survive are those that do.

WHEN TO COMPLAIN

Sometimes, however, dissatisfaction may arise from simply choosing the wrong place for one's needs. Furthermore, some irritation outside the establishment's control may have caused annoyance so that whatever the establishment does to rectify the situation, it may not be to your satisfaction. It is always worth reflecting on these issues before you decide to make a complaint. Clearly if the food or drink is bad, is served at the incorrect temperature, or is not what was ordered, then a complaint is justified. Similarly if the staff are rude, indifferent or unpleasant, then there is also cause for complaint.

— *Legal regulations* —

Within restaurant operations there are also a number of situations that are the subject of *legal regulation*. These are highlighted below. It should, however, be borne in mind that these are highly summarized guidelines and that many of the issues highlighted are affected by the particular circumstances at the time.

Provision of services

Hotel and catering operations are under no obligation to provide services unless the operation is an establishment covered by the *Hotel Proprietors Act, 1956* and the customers seeking services are classed as bona fide travellers. Establishments may refuse to serve people who do not meet the dress requirements of the establishment, eg the wearing of jackets or non-allowance of beachwear. Additionally, licensed establishments may refuse to serve people who are drunk or quarrelsome.

Establishments are, however, under the obligation to ensure that they do not breach the *Sex Discrimination Act, 1975* and the *Race Relations Act, 1976*. These Acts, amongst other things, legislate against discrimination on the grounds of sex, race, creed or colour. Under these Acts, establishments may not refuse services, provide inferior services or set unreasonable conditions on the basis of these characteristics.

Refusing to pay

Under the *Sale of Goods Act, 1979* the customer can refuse to pay for a meal or demand a replacement if:

(a) the goods supplied do not correspond to the description
(b) a displayed item is not what it seems eg a sweet where the cream which would reasonably be expected to be fresh is in fact artificial
(c) the food is inedible or the drink undrinkable.

Additionally, the *Trades Description Acts, 1968/1972* make it a criminal offence to misdescribe goods or services.

Unable to pay

If for some reason you find yourself in the very embarrassing position of being without the means to pay and this is a pure mistake, then the restaurant can seek proof of identity and take your name and address. However, if fraud is suspected, then the police may be called in. The restaurant may not take personal items as security unless you are staying in an hotel covered by the *Hotel Proprietors Act, 1956*. In this case the proprietor has the right of *lien*; in other words, your luggage may be taken pending payment.

Your property and you

If an establishment is covered by the *Hotel Proprietors Act, 1956*, then it is liable for your property while you are staying there. Other than this, there is no automatic liability unless the damage to or loss of your property has resulted from negligence on behalf of the establishment, which would have to be proved. Under the *Health and Safety* legislation (1974 and others) there is a duty on the part of the establishment to care for all lawful visitors, and negligence is a criminal offence. Establishments are, therefore, legally bound to look after your health and safety whilst you are on their premises.

Price lists, service, cover and minimum charges

Restaurants are required to display food and drink prices (*Price Marking Order, 1979*) so they can be seen before entering the premises or, in the case of a complex, before entering the dining area. If service and cover charges are stated on menus and price lists, then they should be paid unless, in the case of service charges, you consider that the service has been poor.

Under the *Food Labelling (Amendment) Regulations, 1989*, there is a requirement that the alcoholic strength (given as a percentage of alcohol by volume) of a representative sample of dispensed drinks be displayed on price lists, wine lists and menus. Exemptions from this requirement are drinks which are less than 1.2% alcohol by volume and cocktails. The number shown need not exceed six in the case of EEC controlled wines or 30 in the case of other alcoholic drinks. Wines sold in bottles are not covered as the alcoholic strength should be shown on the label.

In 1989 Part III of the *Consumer Protection Act, 1987* came into force. This part of the Act deals with misleading prices and, among the provisions, it states that it is an offence to give misleading price information. It recommends that, where the customer has to pay a non-optional charge, these should be incorporated into the total price or they should not be charged at all. It also states that cover charges and minimum charges should be prominently displayed. Compliance with these provisions is not obligatory, but failure to do so could be used as evidence by the Office of Fair Trading that an offence has been committed.

7 Entertaining at home

An important part of our culture involves meeting and getting to know people over a meal. We also celebrate special events, such as birthdays and weddings, in a similar way. Going out for a meal is one way of doing this, the other is to entertain at home.

When we discussed eating out it was suggested that food and drink were not the only important components; this also applies to entertaining at home. Someone once said, 'If you are coming to see me, come any time; if you want to see my home, give me a week's notice'. Therefore, just as it is important to consider your own needs when you go out, it is worth considering the needs of those who are coming to visit you. Why are they coming? Who are they – friends or business acquaintances? What experience of eating and drinking have they had?

It is always best not to overwhelm people whom you do not know well – they might fear inviting you back! Also, if your guests are very experienced and knowledgeable about food and drink, do not assume that simple food is not acceptable. This does not, however, mean that the quality of the food and drink can be poor.

Another very important consideration when entertaining at home is one's own capabilities. As a general rule, never attempt a new dish when you have people round. Working on the premiss that what can go wrong will, experimentation is not advisable, especially when you may be trying to impress. Always keep it simple, bearing in mind that one person's simplicity is another's nightmare. Therefore keep it simple in your own terms. It is better to do a simple meal well than a complex one badly. Overall, try and choose dishes that allow the maximum amount of time to be spent with the guests.

Below are outlined a few pointers on entertaining at home which will help you to consider some of the issues involved.

MENUS

The basic pointers to menu planning are:

1 **Vary colours of food:** this applies to individual courses as well as within courses.
2 **Vary textures:** again between courses and within courses.
3 **Avoid repeating ingredients in different dishes:** eg avoid combinations like mushroom soup followed by chicken with a mushroom sauce. One possible exception to this is fruit where, for instance, melon might be served at the start of the meal and strawberries at the end.
4 **Mix hot and cold dishes:** this not only makes the meal interesting but also helps in terms of preparation and service of dishes. Cold food that has been prepared beforehand prevents the cook from being away from the table too much.
5 **Vary cooking processes:** apart from avoiding, for instance, a meal where all the foods are fried, variations in cooking methods help in the kitchen organization. If some foods are cooked in the oven and others on the hob, the preparations for the meal can be spread out.
6 **Plan a logical menu sequence:** in order to plan the menu, it is worth considering the classical menu sequence detailed on pages 48–50. This sequence assists in the planning of courses so that they follow logically, forming a complete meal.
7 **Consider the ease of eating:** this is particularly important if the meal is to be a buffet – are people to stand up or rest their plates on their laps? Buffets can be:
   a *finger buffets*, where the guests select and consume the food with their fingers
   b *fork buffets*, where the guests select foods which they then eat with only a fork. In this case, the food should be of such a shape and size that this is easily accomplished
   c a *display buffet* from which the guests select their food and then eat at a table.
8 **Keep a record:** if you do a fair amount of entertaining, it is important to keep a record of the guests, what you gave them and how it was received. This avoids the danger of giving people the same meal twice or of providing foods that certain guests may not like.

DRINKS

Over the past few years a significant change that has occurred in the world of eating and drinking has been the increase in consumption of bottled waters. Somehow this has also legitimized the offering of water with meals generally – it used to be extremely difficult to obtain water in a restaurant. With meals at home the same applies, people now expect to be offered water, bottled or otherwise. It is worth pointing out, however, that if mineral water is offered, it should be chilled rather than adding ice made from tap water. Also avoid adding lemon or lime slices to the water as this will affect the taste of the accompanying wine.

Another recent change has been the move towards non-alcoholic and low-alcoholic drinks, not only for those who are driving but also for people who simply prefer not to drink too much. There are now a wide variety of non-alcoholic and low-alcoholic wines and beers, but do try them before serving – some are less than good.

When planning which drinks to serve with a meal, first take into account the notes made on food and wine harmony in Section 4. Additionally, as with food, it is important to bear in mind the guests themselves. If people are not very discerning, then there is really no point in serving expensive wines. Of course, the opposite of this is also true.

Frequently people do not offer a wide range of pre-dinner drinks; light wines or some form of long cocktail such as Kir are commonplace. Sparkling wines are also popular. If you are planning an evening with heavy drinkers and are offering a full range of spirits, etc, then consider making the food simple. Attempts at sophisticated meals are lost on the people who have drunk a fair amount as the palate is desensitized and the ability to appreciate subtlety is lost.

TABLE LAY-UP

The table lay-up should reflect the meal, both in the equipment to be used and the style of the meal. The general convention for laying tables is to lay plates, cutlery and glassware according to the menu and the drinks to be served.

- *Cutlery*: when eating it is usual to start with the cutlery on the outside for the first course, working inwards (if in doubt wait for others or be brave enough to ask)
- *Side knives*: these can be a problem as they may appear in a variety of places: under the sweet cutlery at the top of the place setting; on the side plate; or even as an additional knife laid at the right-hand side of the setting on the inside. It is probably preferable to place this knife on the side plate where it is easier for one's guests to work with
- *Glassware*: this is laid on the right, with the tallest glasses furthest away and the lower ones nearer
- *Napkins*: these are best kept simple and least handled. Elaborate napkin folds are often difficult to unravel and can cause accidents. Overfolded napkins create creases which can look unsightly when the napkin is opened out

Generally it is wise to keep things simple, otherwise there is danger that the décor and lay-up will overshadow the food and drink. It is all a question of balance: elaborate meals call for elaborate décor and lay-ups, with the opposite also being true. It is better to lay a table simply, bringing in additional cutlery, plates and glassware as and when required than to over lay the table. This leads to high expectations in the minds of your guests which you may not be able to meet, thus leading to potential disappointment.

OTHER FEATURES

It is important to remember that one's guests have other needs as well as food and drink. Therefore try to think of other features of the meal that would add to the occasion. These might include such things as small presents for the guests, background music, where after-meal drinks will be served, whether or not an after-lunch walk should be suggested, or perhaps games such as cards or some party games. Also people do like to know how they are expected to dress and whether gifts will be welcomed or frowned upon. Above all, it is best for hosts to be hosts rather than spend all of their time in the kitchen.

Depending on the level of formality it might be worth considering asking your guests to participate in the preparation – although this only really works with people whom you know well. Of course, an alternative is to have someone in to do the food, or to buy it ready prepared from a restaurant or retail store. However, if you consider a caterer, do seek personal recommendations or, at least, try the food before risking it on your guests.

All the above are useful points to consider when entertaining at home. However one must never forget that the best part of most meals is at the end: coffee, drinks, chocolates and conversation.

SECTION

2

On food

1   The menu sequence

Menus may be divided into two classes, traditionally called *à la carte* and *table d'hôte*. The difference between these two is that the à la carte menu has dishes separately priced, whereas the table d'hôte menu has an inclusive price either for the whole meal or for a specified number of courses (eg any two or any four courses).

Over the years the sequence of menus has taken on a classical format or order. This format is used to lay out menus as well as to indicate the order of the various courses. The sequence is as follows:

— 1 Hors-d'oeuvre —

Traditionally this consisted of a variety of salads but now also includes items such as pâtés, mousses, fruit, smoked fish and meats.

— 2 Soups (potages) —

This includes all soups, both hot and cold.

— 3 Egg dishes (oeufs) —

There are a wide variety of egg dishes beyond the usual omelettes but these have not retained their popularity on modern menus.

— 4 Pasta and rice (farineux) —

This includes all pasta and rice dishes.

— 5 Fish (poisson) —

This course consists of fish dishes, both hot and cold. However, such fish dishes as smoked salmon or prawn salads are also considered to be hors-d'oeuvre.

— 6 Entrée —

Usually a small dish of choice cuts complete in itself, eg cutlets, kebabs, steaks and filled vol-au-vent cases.

— 7 Sorbet —

Traditionally sorbets (sometimes called *granites*) were served to give a pause within a meal, allowing the palate to be refreshed. They are lightly-frozen water ices, often based on unsweetened fruit juice, and may be served with a spirit, liqueur or even Champagne poured over. Russian cigarettes also used to be offered at this stage in the meal.

— 8 Relevé —

This refers to the main roasts or larger joints of meat.

— 9 Roast (rôti) —

This traditionally refers to game or poultry.

— 10 Vegetables (légumes) —

Certain vegetables (eg asparagus and artichokes) may be served as a separate course instead of with the main course, in which case they are served at this stage. (The previous dishes would have been garnished with potatoes and vegetables.)

— 11 Salad (salade) —

This refers to a small plate of salad that is taken after the main course (or courses) and is often quite literally a green salad with dressing.

— 12 Cold buffet (buffet froid) —

A variety of cold meats and fish items together with salad garnishes.

— 13 Sweets (entremets) —

This course includes both hot and cold puddings.

— 14 Cheese (fromages) —

— 15 Savoury (savoureux) —

Sometimes simple savouries, such as Welsh rarebit and other items on toast or in pastry or savoury soufflés, are served at this stage of the meal.

— *16 Fruit (dessert)* —

This includes fresh fruit, nuts and perhaps even candied fruits.

— *17 Beverages* —

Traditionally this referred to coffee but nowadays a much wider range of beverages is frequently offered including tea, tisanes and proprietary beverages.

> Note: *Beverages are not counted as a course as such and should therefore not be included when the number of courses is stated. Thus for a formal function, if four courses are quoted, this refers to four* food *courses; the beverages being in addition.*

The sequence outlined above is used throughout the Hotel and Catering Industry when menus are compiled either for à la carte or table d'hôte service. However, with à la carte menus (where the customer chooses), a number of categories are often grouped together. At its most simple a menu might comprise:

- *starters* (groups 1–4)
- *main courses* (groups 5–12)
- *afters* (groups 13–16)
- *beverages* (17).

When planning menus both as a customer in a restaurant and for home entertaining the traditional sequence above is a good guide. Soups tend to work before fish dishes and fish before meat etc. Although the sequence also shows sweets before savouries and cheese, it is now more common in Britain for cheese to be offered *before* the sweet. This allows for the drier main course wine to be consumed with the cheese course, thereby preventing its taste from being destroyed by the change to sweet dishes. Furthermore, changing from savoury to sweet and then back to savoury does not follow a logical pattern of taste sensations and may well affect one's enjoyment of the food. Despite this, either sequence may be found.

2 Food, accompaniments and lay-ups

There are a number of dishes where traditional accompaniments are served. Additionally, tradition has indicated the appropriate lay-up or cover for certain dishes. Below is a guide to these foods, accompaniments and lay-ups. This is not intended to be a prescription as changes are constantly taking place and new accompaniments being tried. Also, the desire for healthier eating has led to alterations: alternatives to butter are often provided, and frequently bread is not buttered in advance, thereby allowing the customer to choose his or her requirements.

Accompaniments are offered with certain dishes mainly to assist in improving the flavour or to counteract richness. For the lay-up the most important consideration is to aid eating. The use of fish knives and forks, for instance, is becoming less fashionable – the original reason for this cutlery revolved around the desire to keep them separate from other knives and forks. In days gone by when silver was less refined and cleaning not so thorough, this was perhaps necessary but it is less true today.

SAUCES

Although there are a wide variety of sauces, they are almost always variations of the same base sauces:

(a) **fond brun** (basic brown sauce) or **demi-glace**
(b) **velouté**
(c) **allemande**
(d) **béchamel**
(e) **tomato sauce**
(f) **mayonnaise** (cold)
(g) **hollandaise** (hot)
(h) **vinaigrette** (cold).

The function of these basic sauces is to supply the fundamental elements for other sauces. That is to say, a variety of different ingredients is added to a given proportion of the basic sauce, giving each sauce its own particular taste. For example:

- An infusion of vinegar and aromatic herbs in a demi-glace sauce constitutes a *poivrade* sauce
- The *aurore* sauce is obtained by adding tomato to a béchamel sauce or velouté
- Béchamel sauce mixed with gruyère or parmesan cheese becomes *Mornay* sauce

Fond brun

Obtained by cooking and rendering down veal bones and sometimes beef bones as well.

Velouté

A golden roux to which a basic white veal, poultry or fish stock is added, depending on what it is to be used for.

Allemande

A velouté thickened with egg yolks and cream.

Béchamel

A white roux diluted with milk.

Tomato

Prepared with fresh or concentrated tomatoes, mirepoix (a vegetable base) and bacon scraps.

Mayonnaise

Consists of egg yolks, oil, vinegar, salt, pepper and mustard.

Hollandaise

A reduction of shallots and vinegar, melted butter and egg yolks.

Vinaigrette

Consists of oil, vinegar, onions, gherkins and chopped herbs.

SIMPLE OR COMPOSITE BUTTERS

The terms *simple butter* and *composite butter* are allocated to:

(a) butter which is simply melted or cooked to varying degrees
(b) butter mixed with certain substances which are normally crushed
into a purée or chopped.

Beurre fondu	(melted butter) butter melted in a double saucepan (porringer)
Beurre noisette	butter cooked until it has a light golden colour
Beurre noir	butter which is cooked until black
Beurre d'anchois	(anchovy butter) butter to which crushed and strained anchovy fillets have been added
Beurre Colbert	*maître d'hôtel* butter with meat glaze and chopped tarragon added
Beurre d'escargots	(snail butter) creamed butter to which crushed garlic, shallots and chopped parsley have been added
Beurre de homard	(lobster butter) creamed butter to which eggs and crushed and strained lobster flesh have been added
Beurre maître d'hôtel	creamed butter mixed with chopped parsley, lemon juice, salt, pepper and Worcester sauce
Beurre moutarde	(mustard butter) butter which has been softened and mixed with mustard

HORS-D'OEUVRE

Traditionally eaten with a fish knife and fork, the cutlery is nowadays more likely to be dictated by the type of food and its presentation. Oil and vinegar were also traditionally offered but this has become less common becacuse the salads themselves are already well-dressed. Brown bread and butter were often offered but this is now less common. Hors-d'oeuvre items include salads, fish, meats, canapés and eggs.

Salads

Plain or compound. Examples include Russian salad, fish salad, game or poultry salad, ox tongue salad, cucumber salad, potato salad, tomato salad, beetroot salad, white celery, grated celery, red cabbage, cauliflower, tomatoes Antiboise, Andalousian, Beaulieu, Monaco, Parisienne, Printanière etc. (For a selection of compound salads see pages 73–5.)

Fish

- *Anchovy*: raw, filleted, marinated, sliced, with peppers, stuffed paupiettes, Toulonnaise and Nîçoise
- *Herring*: fresh or marinated
- *Lobster*: half lobster, lobster pieces, lobster salad
- *Mackerel*: small mackerel, marinated or in oil
- *Oysters*: raw with lemon or oyster cocktail
- *Prawns*: in cocktail, stuffed, marinated or in mousse
- *Sardines*: in oil or in tomato sauce
- *Sea food*: a variety of edible marine shell fish
- *Shrimps*: pink or grey, boiled, unopened or fully prepared
- *Smoked eel*: filleted or sliced
- *Smoked salmon*: in thin slices rolled into cornet shapes
- *Trout*: grilled, in aspic
- *Tunny*: in oil, in purée or marinated

Meats

- *Dried meats*: (grisons) sliced very thin
- *Foie gras*: in slices, in aspic, in mousse, in terrine, in individual pâtés, etc
- *Ham*: raw, boiled, smoked, in slices, in rolls
- *Pâtés*: miscellaneous, sliced
- *Salami*: of all sorts, sliced thin
- *Sausage*: foie gras, pheasant, chicken etc

Canapés

These are slices of bread with the crust removed, cut into a variety of shapes, then toasted, fried in oil or butter and garnished.

Canapés are garnished with caviar, smoked salmon, foie gras, anchovies, tunny, shrimps, prawns, eggs, raw and cooked ham, veal, tongue, salami, game, Gruyere cheese, asparagus tips etc.

Eggs

Poached or in aspic, hard boiled and cut in two, garnished or stuffed, with spinach, macédoine, Strasbourg, Riga, Toulonnaise, with mayonnaise, salad, vinaigrette etc.

OTHER STARTERS
— *Fruit cocktails* —

Usually these are eaten with a teaspoon and sugar is offered where there is grapefruit in the cocktail.

— *Fruit juices* —

Sometimes castor sugar may be offered in which case a teaspoon should also be given to stir in the sugar. For tomato juice, salt and Worcester sauce are offered and again a teaspoon given to aid mixing in these accompaniments.

— *Fresh fruit* —

Usually no accompaniment is offered although some people might like castor sugar. Both castor sugar and ground ginger should be offered with melon if it is served by itself.

— *Oysters* (huîtres) —

Cold oysters are usually served in one half of the shell on a bed of crushed ice in a soup plate. An oyster fork is used but a small sweet fork may suffice. Because the oysters are eaten holding the shell in one hand, a finger bowl could be offered. Accompaniments offered include cayenne pepper, a peppermill, chilli vinegar, and Tobasco sauce – these four condiments being collectively known as an *oyster cruet*. Additionally half a lemon and brown bread are usually given.

— *Snails* (escargots) —

Snail tongs should be placed on the left and a snail fork on the right. The snails themselves are served in an escargot dish which has six or twelve indentations. French bread is offered for mopping up the sauce. Half a lemon is also given and a finger bowl could be offered.

— *Potted shrimps* —

A fish knife and fork or a small knife and fork should be laid. Accompaniments include hot, unbuttered breakfast toast (there is plenty of butter already in this dish and brown bread and butter would not be appropriate), cayenne pepper, a peppermill, and segments of lemon.

— *Mousses and pâtés* —

Normally these are eaten with hot breakfast toast using a side knife and sweet fork. Butter is also offered. The garnishes served should be appropriate to the dish itself, eg lemon with fish mousses – although lemon is often also offered with pâtés.

— *Smoked salmon* (saumon fumé) —

The traditional accompaniments are cayenne pepper, a peppermill, segments of lemon and brown bread and butter. An alternative to brown bread and butter could be unbuttered rye or granary bread, allowing the customers to help themselves to butter or alternatives. Oil is sometimes offered.

— *Other smoked fish* —

As well as the accompaniments offered with smoked salmon, horseradish sauce has become a standard offering with smoked fish. This, however, is often served straight from the bottle and really is far too strong for the fish. An alternative is to mix half horseradish with acidulated cream or natural yogurt, a little lemon juice, some chopped parsley and fine strips (julienne) of lemon zest. This makes a softer and more pleasant accompaniment for these fish dishes.

— *Asparagus* —

When served hot, asparagus can be accompanied by either melted butter or hollandaise sauce. It is useful to put a fork under the plate in order to tilt the plate to the left so that the butter or sauce will form in a well at the bottom of the plate. The asparagus can be eaten with the hands or with holders. A finger bowl and a spare napkin should be offered.

Asparagus can be served cold with a vinaigrette dressing and eaten with a side knife and sweet fork. Cold asparagus is usually served as a starter.

— *Corn on the cob* —

Corn cobs are usually served with special holders which are like small swords or forks. Three wooden cocktail sticks in each end will also do the job, but avoid trying to use two sweet forks as it is possible to painfully catch teeth on the prongs. There are special dishes available, but a soup plate will do in order to provide a reservoir for the melted butter or sometimes hollandaise sauce. A finger bowl and a spare napkin might be advisable. A peppermill is offered.

— *Globe artichoke* —

This vegetable is usually served whole as a starter. The edible portion of the leaves is 'sucked off' between the teeth after dipping them in a dressing. The leaves are held with the fingers. The heart is finally eaten with a side knife and sweet fork. A finger bowl and a spare napkin are essential. There are special dishes for this vegetable, but a fish plate and a small bowl to hold the dressing will do just as well.

SOUPS

Soups are divided up into several different categories:

(a) **consommés**   (d) **purées**
(b) **veloutés**   (e) **potages**
(c) **crèmes**   (f) **various national soups.**

In addition, there are bisques (shell fish soups) and broths, eg beef, chicken and vegetable.

— *Consommés* —

A consommé is made from a poultry, beef, game, or vegetable bouillon, which is subjected to clarification (minced meat mixed with vegetables and other flavours), cooked and passed through a fine cloth. It can be served as it is, or chilled.

These soups are usually served in consommé cups with a sweet spoon given. Traditionally this soup was eaten before going home after a party either in a carriage or on horse. It was therefore eaten standing up and the spoon employed merely to help in eating the garnish. This tradition has been followed through to the table, but it would be a brave person who picked up the cup and drank from it, even though this is what the handles were for!

— *Veloutés, crèmes and purées* —

Veloutés

These are based either on a poultry, basic white or fish stock. Normally they consist of a pale roux mixed into a basic stock. They are thickened with cream and egg yolk and then have a little butter added to them at the last moment.

Cream soups (*les crèmes*)

These differ from veloutés in that milk is stirred into them in place of some of the cooking stock and they are not thickened. There are no special accompaniments for most of the cream soups. Croûtons (fried, small cubes of bread) are offered traditionally only with tomato soup but it is now more common to expect these with any soup.

Purées

These have a vegetable purée base and are thickened by the natural starch content of the ingredients. There are some basic purées, eg potato (*Parmentier*), and dried green pea (*Saint-Germain*) and these are the basis of many other derived types of purée. The derived purées are simply obtained by adding a further ingredient which then characterizes the soup and gives it its name. Croûtons are offered with most purée soups.

— Potages —

This description is given to *unstrained* soups of a rather country style, the main types of which are cabbage soup and vegetable soups, which are generally quite thick.

— National soups —

Batwinia	(Russian) purée of spinach, sorrel, beetroot and white wine, with small ice cubes served separately. Served very cold
Beer	(German) beer velouté with cream and powdered cinnamon, poured over toast in a soup dish
Bortsch	(Polish) duck flavoured consommé garnished with duck, diced beef and turned vegetables; the accompaniments are sour cream, beetroot juice and bouchées filled with duck paste. A soup plate rather than a consommé cup should be used owing to the large amount of garnish
Busecca	(Italian) powdered white of leek, crushed tomatoes, haricot beans and veal membrane
Bouillabaisse	(French) this soup is really a fish stew. Although a soup plate and soup spoon should be used, a side knife and sweet fork should also be given. Thin slices of French bread, dipped in oil and grilled, should also be offered with this dish
Cherry	(German) bouillon consisting of purée, cherry juice, and red wine garnished with stoned cherries and sponge finger biscuits
Chicken broth	(English) chicken broth, garnished with rice
Cock-a-leekie	(Scottish) veal and chicken consommé garnished with shredded leeks, chicken and prunes

Cuzido	(Portuguese) clear stew made from beef, chicken, ham, smoked sausages, cabbage and rice
Kroupnich	(Russian) barley and sections of poultry offal garnished with small vol-au-vents stuffed with poultry meat
Mille fanti	(Italian) consommé with a covering of bread crumbs, Parmesan and beaten eggs
Minestroni	(Italian) vegetable *paysanne* soup with crushed tomatoes and spaghetti, thickened with bacon fat, garlic and chopped parsley. The traditional accompaniments for this soup are grated Parmesan cheese and grilled flutes
Miss Betsy	(English) consommé of pearl barley, celery and tomato purée, garnished with apples
Mulligatawny	(English) poultry velouté with curry, garnished with rice and shredded chicken
Mutton broth	(English) mutton bouillon with pearl barley, coarsely chopped vegetables and diced mutton
Olla-podrida	(Spanish) broth made from oxtail, pork, mutton, ham, sausage, chicken and chick peas
Oxtail, clear	(English) oxtail consommé garnished with oxtail and coarsely sliced vegetables, flavoured with Madeira
Oxtail, thick	(English) as above but thickened with a brown roux
Petit marmite	(French) beef and chicken flavoured soup garnished with turned root vegetables and dice of beef and chicken. This soup is served in a special marmite pot which resembles a small casserole. A sweet rather than a soup spoon is often provided, as it is easier to remove the soup from the pot with the former (because of the width of the top of the pot). Accompaniments are grated Parmesan cheese, grilled flutes and poached marrow bone

Puchero	(Spanish) the same as *Olla-podrida* but more heavily garnished
Shchy	(Russian) *bortsch* consommé, garnished with sauerkraut and served, separately, with beetroot juice and sour cream
Terrapin	(American) clear or thickened turtle soup, garnished with diced terrapin meat
Turtle, clear	(English) beef, poultry and turtle consommé with a strongly aromatic herb flavouring, garnished with diced turtle meat. This soup should be served in a consommé cup with cheese straws, lemon segments, and brown bread and butter. A measure of warmed sherry or sometimes Madeira should also be put into the soup just before serving
Turtle, thickened	(English) the same as clear turtle soup but thickened with a brown roux
Zuppa pavese	(Italian) consommé containing a whole egg and toast, covered with cheese and grilled

EGGS

There are several hundred different ways of preparing eggs, but only a few methods of cooking them. These are listed below:

(a) **boiled**
(b) **hard boiled** (*dur*)
(c) **soft boiled** (*à la coque*)
(d) **scrambled** (*brouillé*)
(e) **coddled** (*en cocotte*)
(f) **poached** (*poché*)
(g) **fried** (*frit*)
(h) **fried served on the dish** (*sur le plat*)
(i) **omelette.**

Egg dishes as separate courses have a chequered history. Omelettes have retained their popularity, but dishes such as eggs en cocotte (baked in a small earthenware dish) occasionally feature on the menu.

PASTA, RICE AND
CEREAL BASED FOODS

— *Pasta* (macaroni, spaghetti, noodles, — cornettes etc)

Pasta are cooked in boiling salted water for between 12 and 16 minutes. For spaghetti, a joint fork should be laid on the right and a sweet spoon on the left. For all other dishes these should be eaten with a sweet spoon and fork laid in the usual way. Grated Parmesan cheese is normally offered with these dishes.

— *Rice* —

À l'Indienne	cooked in boiling water, cooled then reheated in a heavily buttered dish. It is served very dry
Pilaf(f)	prepared with onions and butter and cooked in the oven with a little bouillon so that it is quite dry and the grains separate easily
Risotto	prepared with onions, butter and a bouillon, but this time more copiously damped with the bouillon. When it is finished, it is bound with butter and grated Parmesan
Milanaise	a saffron rice risotto to which mushrooms and small squares of fresh tomato have been added

— *Ravioli* —

Ravioli takes the form of small squares of ravioli dough (a type of noodle dough) filled with the same type of stuffing as is used for cannelloni. They are poached in salted water, placed in an oven dish in alternating layers (first a layer of ravioli, then a layer of tomato, veal and grated cheese and so on), then covered with melted butter and baked or glazed in the oven.

— *Cannelloni* —

Small rolls of pasta filled with minced poultry meat, but more frequently with minced meat and spinach, placed in the oven and baked, covered with a little tomato sauce and grated cheese.

— *Gnocchi* —

Parisienne	made from a light choux paste prepared from flour, butter, eggs and milk. Formed into small round plugs by means of a special case, boiled in water and placed in a hollow dish. Covered with a béchamel sauce and grated cheese. They are baked in a slow oven so that they can both open out and become golden at the same time. To be served quickly as for a soufflé
Piémontaise	made from a dough consisting of ⅔ potatoes and ⅓ flour, together with some eggs. Rolled into small balls and then poached in salted water and placed on an oven dish. They are then covered with a tomato sauce or a tomato flavoured Madeira sauce, coated with cheese and lightly browned off in the oven
Romaine	prepared from a basis consisting of semolina, milk, eggs and cheese, cooked and allowed to cool, cut into various shapes, baked in the oven and served with a Madeira sauce with a heavy tomato content

FISH

The most well-known types of fish and the more commonly used methods of preparing and garnishing them are given below:

— *Fresh water fish* —

Barbel	Burbot	Chub
Bream	Carp	Dace
Brown trout	Char	Eel

Farmed trout
Grayling
Gudgeon
Lake trout

Perch
Pike
Pike perch
Pollan

Salmon
Salmon trout
Sturgeon
Tench

— *Salt water fish* —

Anchovy
Bass
Brill
Cod
Conger
Dorade
Flounder
Haddock
Hake

Halibut
Herring
Ling
Mackerel
Mullet
Plaice
Red mullet
Saithe
Salt cod

Sardine
Skate
Smelt
Sole
Tunny
Turbot
Whiting

— *Shellfish and molluscs* —

Coquille-
   Saint-Jacques
   (scallops)
Crab
Crayfish

Lobster
Mussels
Oysters
Prawn
Scampi

Shrimp
Snails
Turtle

— *Preparation* —

Fried
Small fish or large fish cut up into fillets or slices, dipped in milk and flour and fried in very hot deep oil. Generally accompanied by a cold sauce or lemon and fried parsley. (Never cover fried fish as the crispness will soon disappear.)

Grilled
Seasoned, dipped in oil and cooked on a very hot grill, accompanied by a sauce or a composite butter.

Matelote
The same method as for fish *au vin blanc* but the white wine is replaced by a good quality red wine.

Meunière
Fried in butter in a pan, dressed on the dish with lemon juice and chopped parsley and covered with hot melted butter.

Orly
Cut into fillets or slices, dipped in batter and deep fried. Accompanied with tomato sauce or a cold sauce.

Poached (hot)
Cooked in an aromatic stock. Generally accompanied with hollandaise sauce, mousseline or melted butter.

Poached (cold)
The same method as above but allowed to cool in the stock. Accompanied with an attractive garnishing and cold sauce.

Vin blanc (au)
Poached in white wine. Covered with a white wine sauce to which is added the reduced stock from the cooking.

— *Accompaniments for fish* —

- For hot dishes served with a sauce there is not usually any additional accompaniment
- Hot dishes without other sauces often have hollandaise sauce offered
- For fried fish which has been breadcrumbed (*à l'Anglaise*), tartare sauce (or another mayonnaise-based sauce) is normally offered with segments of lemon already on the plate
- For fried fish, not breadcrumbed, or grilled fish there are no special accompaniments other than providing lemon. Sometimes sauces such as hollandaise or tartare are offered
- For fish which has been battered and deep fried (*à l'Orly*), a tomato sauce is sometimes offered
- Mayonnaise or another similar mayonnaise-based sauce such as Sauce Vert are usually offered with cold poached fish

MEATS

— *Roast meats* —

In all cases roast gravy is offered. For dishes where the roast is plain (ie not roasted with herbs for instance) the following are usually offered:

- *Roast beef* – horseradish sauce, mustards and Yorkshire pudding
- *Roast lamb* – mint sauce and also more commonly redcurrant jelly
- *Roast mutton* – redcurrant jelly and sometimes white onion sauce
- *Roast pork* – apple sauce and sage and onion stuffing

— *Boiled meats* —

- *Boiled mutton* – caper sauce
- *Salt beef* – turned root vegetables, dumplings and the natural cooking liquor
- *Boiled fresh beef* – turned root vegetables, natural cooking liquor, rock salt and gherkins
- *Boiled ham* – parsley sauce or white onion sauce

— *Irish stew* —

The stew is often served in a soup plate with Worcester sauce and pickled red cabbage as accompaniments.

— *Curry* —

General accompaniments are poppadums (crisp, highly-seasoned pancakes) and Bombay Duck (dried fillet of fish from the Indian Ocean) and mango chutney.

— *Mixed grill and other grills* —

Various mustards and sometimes proprietary sauces act as accompaniments. These dishes are usually garnished with cress and sometimes potato.

— Roast chicken (poulet rôti) —

The accompaniments should be bread sauce, roast gravy and parsley and thyme stuffing.

— Roast duck (caneton rôti) —

Sage and onion stuffing, apple sauce and roast gravy should be served.

— Wild duck (canard sauvage) —

Roast gravy should accompany the meat and an orange salad with an acidulated cream dressing offered as a side dish.

— Roast goose (oie rôti) —

Sage and onion stuffing, apple sauce and roast gravy should act as accompaniments.

— Roast turkey (dinde rôti) —

Cranberry sauce, chestnut stuffing, chipolata sausages, game chips, watercress and roast gravy are the usual accompaniments.

— Hare (lièvre) —

Heart-shaped croûtons, forcemeat balls and redcurrant jelly should act as accompaniments.

— Venison (venaison) —

Cumberland sauce and redcurrant jelly are the usual accompaniments.

— All feathered game —

Fried breadcrumbs, hot liver pâté spread on a croûte on which the meat sits, bread sauce, game chips, watercress and roast gravy may all be served.

NOTES ON CARVING

The carving of meats is a skill that is developed only through practice. Two of the most important things in carving are a sharp knife (blunt ones cause accidents) and confidence. Some general points to consider are:

- Meat is generally carved across the grain
- Always use a very sharp knife. For most joints use a knife with a blade 25–30 cm (10–12 in) long and about 2.5 cm (1 in) wide; for poultry or game it is best to use a knife with a blade 20 cm (8 in) long; for ham a long, flexible carving knife is used – often called a ham knife. Serrated knives do not always cut better than plain bladed knives – the latter gives a cleaner cut
- When carving or jointing, the meat should be held firmly, usually with a carving fork, in order to avoid accidents
- Carve on a board (either wooden or plastic). Avoid carving on metal. Apart from the damage this can cause, especially to silver, small splinters of metal can become attached to the meat slices

Methods of carving:
- Chicken – medium-sized birds are often dissected into eight pieces
- Duck – usually dissected into four portions: two legs, two wings and the breast cut into long strips down the breast.
- Poussin and small feathered game – either serve whole or split into two portions
- Beef and ham – slice very thinly
- Lamb, mutton, pork, tongue and veal – usually sliced at double the thickness of beef and ham
- Boiled beef and pressed meats – generally carved slightly thicker than roast meats with boiled beef carved with the grain to avoid the meat shredding

— *Carving techniques* —

Chicken

Lay the chicken on its side on the board. Hold the bird firmly with the flat of the knife and slide the fork beneath the leg joint and raise the leg until the skin is taut. Cut round the taut skin and pull the leg away from the main joint. Cut the leg into two pieces – the drumstick and thigh – and remove the claw end from the drumstick. Turn the bird over and repeat this process with the other leg.

Turn the bird onto its back and holding the main joint firm with the fork, carve down through the wing joint taking part of the breast portion with it. Turn the bird around and repeat with the other breast. Cut down the breast on each side and lever off each breast in turn.

This makes eight pieces. When serving, serve a breast portion with a drumstick and a wing portion with a thigh.

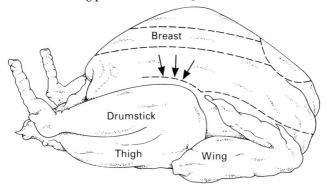

Note: Having completed the carving of the chicken, the carcase should be turned over. The 'oyster' piece is found on the underside of the carcase and is a small, brown portion of meat found on either side of the back.

Duck

The joints on a duck are much tighter and more compact than on a chicken, and the wing joints lie a little further under the base of the carcase than those on a chicken.

The legs and wing joints are removed in a similar way to that described for the chicken, although they are left as whole portions. The breast is then removed and sliced down the length as indicated in the diagram at the top of page 70. If carving the breast on the bone, make the cuts at a slight angle so as to avoid blunting the knife but still cutting straight down onto the bone.

For duckling, the wing portion is not usually separated from the breast.

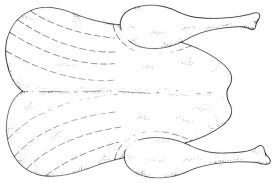

Turkey

The legs and wings should be separated but not entirely removed. This enables carving to be carried out both on the breast and the legs and wings, so that each person can have a portion of dark and a portion of white meat. Dark meat is often put on the plate first, and then the white meat partially covering it. An alternative is to separate the legs and wings from the bird, and to carve these separately followed by the breast.

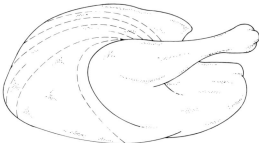

Roast sirloin of beef – on the bone

This joint comprises two main parts, the undercut or fillet, and the uppercut or sirloin. Carving for each part is by slicing down towards the bone – the base of the slice is released by running the knife along the bone underneath the cut slice. Ribs of beef are carved in the same way. The sizes of slice are a matter of personal choice, with some people preferring fairly thick slices and others preferring their beef really thin.

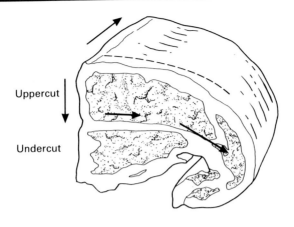

Uppercut

Undercut

Feathered game

For a 'large' bird such as partridge, three portions can be obtained as shown in the illustration. Depending on size, other feathered game can either be served whole or split through the middle of the breast bone.

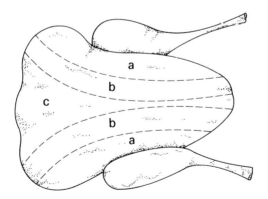

portion a   one leg and one wing with a little of the breast attached,
portion b   as for a
portion c   the breast left on the bone. If large it could be split into two portions by carving down through the breastbone.

Roast best end of lamb

This joint is usually carved by holding the joint upright and slicing down between the cutlet bones, using them as a guide. An alternative is to lay the joint down with the fat uppermost before carving through the fat and the meat, again using the cutlet bone as a guide.

Roast loin of lamb or pork

The loin of lamb or pork may be boned, rolled, stuffed and roasted, or roasted on the bone. In the latter case, the joint is carved into fairly large pieces (the thicknesses of chops) by cutting between the ribs.

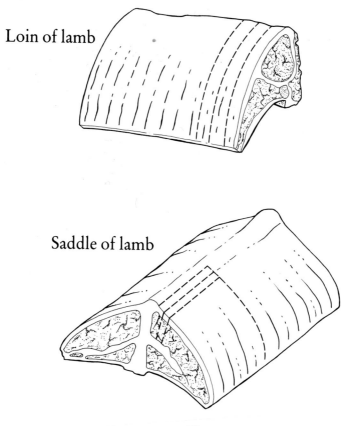

Loin of lamb

Saddle of lamb

A saddle of lamb is a cut of mutton consisting of both loins. Carving can either be by removing each loin from the bone and then carving, or by carving in a similar fashion to ribs of beef or by carving down the length of the meat as in the illustration.

Leg of lamb or pork

Initial carving is to the bone. First, a V-shaped slice is removed from the joint as shown in the illustration. Slices are then carved from the joint by cutting towards the bone outwards from the first cut. The underside is then carved in the same way. For pork, the slices should be fairly thin, and for lamb they should be cut more generously.

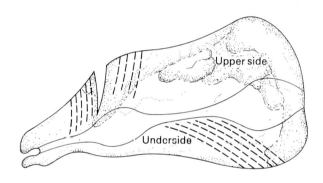

COMPOUND SALADS

The following is a list of compound salads that are offered as part of an hors-d'oeuvre selection, or as part of a cold table buffet.

Aïda	with endives, artichoke hearts, pimentoes, tomatoes, the whites of hard boiled eggs and a mustard vinaigrette
Américaine	with potatoes, tomatoes, celery, onions, hard boiled eggs and vinaigrette
Andalouse	celery, onions, peppers, tomatoes, rice and vinaigrette

Archiduc	beetroot, chicory, truffles, potatoes and vinaigrette
Argenteuil	mixed diced vegetables, covered with asparagus tips and sliced hard boiled egg; mayonnaise
Béatrix	poultry breasts, truffles, asparagus tips, potatoes and a light mayonnaise
Caprice	shredded tongue, ham, poultry, truffles, chicory and vinaigrette
Carmen	red peppers, poultry breasts, peas, rice; vinaigrette with chopped tarragon
Châtelaine	hard boiled eggs, truffles, artichoke hearts, potatoes, vinaigrette with chopped tarragon
Cressonière	potatoes, cress, hard boiled eggs and chopped parsley
Demi-deuil	truffles, potatoes; mayonnaise
Florida	lettuce, quartered oranges, cream and lemon juice
Grande-duchesse	truffles, potatoes, celery, mushrooms; mayonnaise
Italienne	vegetable salad, cubes of salami and anchovy fillets; mayonnaise
Japonaise	tomatoes, pineapple, oranges, lettuce hearts; lemon juice and cream
Jockey-Club	asparagus tips, truffles; mayonnaise
Lorette	mace, celery, beetroot; vinaigrette
Mascotte	asparagus tips, hard boiled eggs, peeled shrimps, truffles, mustard and cream
Mignonnettes	a large brunoise chopped vegetable salad, mayonnaise
Mimosa	lettuce hearts, oranges, grapes, bananas, cream and lemon juice
Monégasque	potatoes, quarter artichokes, tomatoes, black olives, mustard sauce and anchovies
Niçoise	French beans, tomatoes, potatoes, anchovy fillets, tuna, olives and capers; vinaigrette
Opéra	chicken breasts, celery, truffles, tongue, asparagus tips, gherkins, light mayonnaise
Parisienne	slices of crayfish, truffles, Russian salad, bound with mayonnaise and aspic

Rachel	celery, truffles, artichoke hearts, asparagus tips, potatoes, light mayonnaise
Russe (Russian)	carrots, turnips, peas, potatoes, French beans; mayonnaise
Tourangelle	potatoes, French beans, broad beans, mayonnaise made with cream and chopped tarragon
Tosca	poultry, truffles, celery, mayonnaise
Waldorf	apples, celery, nuts, mayonnaise

CHEESE

Cheese is usually served with biscuits but also sometimes with various breads. Butter or an alternative is often expected by guests. Radishes and celery may also be offered as may fresh fruit such as grapes, apples or pears. Salt is provided for the celery and castor sugar for cream cheeses. The lay-up should be either a side knife only or a side knife and sweet fork. The latter is employed where the cheese is to be eaten from the plate rather than being put on a biscuit or bread first.

For best results the cheese should be allowed to temperature by being left out of the fridge for at least one hour before serving. Cheese can be stored in a cool place such as a cellar or fridge but should be wrapped in aluminium foil to prevent it from drying out.

Cheeses are distinguished by flavour and categorized according to their texture. They differ from each other for a number of reasons, mainly arising through variations in the making process. Differences occur in the rind and how it is formed, in the paste, and in the cooking (both time and temperature). Additionally cheeses vary because the milk used comes from different animals: cow, sheep, or goat.

The texture of a cheese depends largely on the period of maturation, the main categories being:

(a) **fresh**
(b) **soft**
(c) **semi-hard**
(d) **hard.**

An additional category is often identified for the various **blue** cheeses.

— *Fresh cheese* —

Cottage	unripened low fat, skimmed milk cheese with a granular curd. Originated in the United States and now has many variations
Cream	similar to *cottage cheese* but is made with full milk. There are a number of different varieties available, some made from non-cow milks
Mozzarella	Italian cheese made now from cow's milk but originally from buffalo milk
Quark	continental version of *cottage cheese*, sharper and more acidic than the American versions. Mainly German and Austrian varieties available
Ricotta	Italian cheese made from the whey of cow's milk. A number of other Italian varieties are available made from sheep's milk

— *Soft cheese* —

Barbery, Fromage de Troyes	French cheese similar to *Camembert* but sometimes sold unripened as a fresh cheese
Bel Paese	this light and creamy Italian cheese has a name which means 'beautiful country' and was first produced in 1929
Brie	famous French cheese made since the eighth century. Other countries now make this style of cheese, distinguishing it by the country's name, eg *German Brie*
Camembert	famous French cheese which is stronger and can be more pungent than *Brie*
Chalichou	French cheese made from goat's milk and coming from Poitou
Epoisses	French Burgundian cheese, speciality of the area. Sometimes dipped in Marc de Bourgogne (local brandy) before being sold

Feta	Greek cheese made from both goat's and sheep's milk
Liptauer	Hungarian cheese spread made from sheep's and cow's milk. Often found with various additions, eg onions, mustard or spices
Munster	French Vosges cheese similar to *Camembert* in shape but with an orange red rind. American, German and Swiss versions are also available
Straccihino	Italian cheese originally from Lombardy. A soft, delicate cheese which now has a number of varieties

— *Semi-hard cheese* —

Appenzeller	typical example of Swiss cheese textures. The name is from the Latin for 'abbot's cell'
Brick	tangy American cheese which is so called because of its shape and the colour of the rind. It is yellow-white in colour and contains small holes. Sadly most American cheese is inferior to the richness of the varieties from Britain and the rest of Europe
Caerphilly	buttermilk-flavoured cheese with a soft taste. Some people find it almost soapy. Originally a Welsh cheese but now manufactured all over Britain
Cantal	French cheese from the Auvergne, similar to *Cheddar*
Cheddar	classic British cheese now made all over the world and referred to as, for example, *Scottish Cheddar, Canadian Cheddar* etc
Cheshire	crumbly, slightly salty cheese, available as either white or red. It was originally made during the twelfth century in Cheshire but is now made all over Britain
Chèvre	the name means 'goat', which denotes the

	origin of the milk from which this cheese, and the wide variety of variations, is made
Colby	another American cheese similar to *Cheddar*
Danbo	square Danish cheese related to *Emmenthal*; sometimes flavoured with caraway
Derby	English Derbyshire cheese now more often known by the sage-flavoured variety, *Sage Derby*
Dunlop	this fairly bland Scottish cheese is similar to *Cheddar* and is said to have come from an Irish recipe
Edam	similar to, but harder than, *Gouda* this Dutch cheese has a fairly bland, buttery taste and a yellow or red wax coated rind. It is sometimes flavoured with cumin
Emmenthal	the name of this Swiss cheese refers to the Emme Valley. It is similar to *Gruyère*, although it is softer and slightly less tasty
Esrom	similar to the French *Port Salut*, this Danish cheese has a red rather than yellow rind
Gjetost	originally goat (*gje*) cheese (*tost*), now mainly found as a cow's milk cheese. It has a firm consistency and a sweetish taste
Gloucester/Double Gloucester	full-cream, classic English cheeses originally made only from the milk of Gloucestershire cows
Gouda	buttery textured, soft and mild flavoured well-known Dutch cheese with a yellow or red rind
Gruyère	mainly known as a Swiss cheese but both the French and Swiss varieties can legally be called by this name. It has small pea size holes and a smooth relatively hard texture. The French varieties may have larger holes
Havarti	named after the farm of the developer (Mrs Hanne Neilson), this Danish cheese is similar to *Tilsit* and has a sourish, slightly sharp taste
Jarlsberg	similar to *Emmenthal*, this Norwegian cheese was first produced in the late 1950s. It has a yellow wax coating

Lancashire	another classic English cheese similar to *Cheshire*; (white *Cheshire* is sometimes sold as *Lancashire*)
Leicester	mild flavoured and orange coloured English cheese
Limberger	often quite pungent, this originally Belgian cheese is now also available from Germany
Manchego	relatively hard cheese which may have holes and has either a white or sometimes yellow paste. Made in Spain from sheep's milk
Monterey	creamy, soft American cheese with many holes. A harder version known as *Monterey Jack* is suitable for grating
Mysost	similar to *Gjetost*, this Norwegian cheese has a firm consistency and a sweetish taste. Now made from cow's milk
Pont l'Evêque	similar to *Camembert*, but square in shape, this French cheese originates from Normandy
Port Salut	mild flavoured cheese with a name meaning 'Port of Salvation', referring to the abbey where exiled Trappist monks returned after the French Revolution
Reblochon	creamy, mild flavoured cheese from the Haute-Savoie region of France. The name comes from the illegal 'second milking' from which the cheese was originally made
Samso	originally made on the Danish island of the same name as a copy of the Swiss *Emmenthal* cheese. There are now many variations
Sauermilchkäse	fairly sourish, sharp German cheeses with a colour ranging from pale yellow to red-brown. The name refers to a family of cheeses available under a variety of names
Tilsit	strong flavoured cheese from the East German town of the same name where it was first produced by Dutch living there. Now available from other parts of Germany
Wensleydale	Yorkshire cheese originally made from sheep's or goat's milk but now from cow's

milk. This cheese is the traditional accompaniment to apple pie

— *Hard cheese* —

Asiago d'Allevio	piquant grating cheese from Vicenza in the north-west of Italy
Asiago Grasso do Monte	similar to *Asiado d'Allevio*, but milder tasting with a smoother paste
Bergkäse	smooth, mild flavoured cheese from Austria
Caciocavallo	originating from ancient Roman times, the name means 'cheese on horseback' because its shape is said to resemble saddlebags
Kefalotyri	literally Greek for 'hard cheese', this is a tasty, grating-type cheese from Greece
Parmesan	classic Italian hard cheese, more correctly called *Parmigiano Reggiano*, and predominantly known as the grated cheese used in and for sprinkling over Italian dishes
Pecorino	hard, sheep's milk, grating or table cheese from southern Italy. Also available with added peppercorns as *Pecorino Pepato* from Sicily
Provolone	smoked cheese made in America, Australia and Italy. Now made from cow's milk but originally from buffalo milk. Younger versions are softer and milder than the longer kept varieties
Sapago	green coloured cheese with an aroma of dried clover. It is a hard, grating-type cheese from Switzerland

— *Blue cheese* —

Bavarian Blue	rich, high fat and sourish German cheese
Blue d'Auvergne	French, Auvergne cheese which becomes blue through the addition of *Penicillium glaucum* mould

Blue de Bresse	fairly soft and mild flavoured French cheese from the area between Soane-et-Loire and the Jura
Blue de Laqueuille	mellow flavoured French cheese from Puy-de-Dome
Blue Cheshire	one of the finest of the blue cheeses which only becomes blue accidentally, although the makers endeavour to assist this process by pricking the cheese and maturing it in a favourable atmosphere
Blue Shropshire	similar to *Blue Cheshire* but made in Scotland
Danish Blue	one of the most well-known of the blue cheeses. Softish and mild flavoured, it was one of the first European blue cheeses to gain popularity in Britain
Dolcelatte	factory made version of *Gorgonzola*. The name is Italian for 'sweet milk' and the cheese is fairly soft with a creamy texture and greenish veining
Gorgonzola	softish, sharp flavoured, classic Italian cheese with greenish veining, which is developed with the addition of mould culture
Roquefort	classic, sheep's milk cheese from the southern Massif Central in France. The maturing takes place in caves which provide a unique humid environment which contributes to the development of the veining
Stilton	famous and classic English cheese made from cow's milk; so called because it was noted as being sold in the Bell Inn at Stilton by travellers stopping there. According to legend it was first made by a Mrs Paulet of Melton Mowbray. Traditionally served by the spoonful but nowadays usually (and perhaps preferably) portioned. The pouring of port on to the top of a whole *Stilton*, once the top rind had been removed, was also popular but this practice is also on the

decline. The *white Stilton* has also become popular and is slightly less flavoursome than the blue variety

SWEETS

The range of possible sweets is very extensive and varied. It can include:

- Cold mousses in a variety of fruit flavours
- Fritters (*beignets*)
- Various crèmes including blancmange and syllabubs charlottes either hot or cold
- A variety of coupes
- Ices
- Omelettes with a variety of fillings and flavourings, eg rum, jam, or apple
- Pancakes with a variety of fillings, eg cherries or other fruits
- Puddings including bread and butter, cabinet, diplomate and various fruit puddings and baked apples
- Soufflés
- Fruit salads
- Gâteaux
- Pies, flans and other pastries

Recently there has been a welcome return to the various traditional sweets, especially puddings such as bread and butter and summer pudding.

There are no particular accompaniments to sweets and the choice of whether to serve on a plate or in a bowl is dependent on the texture of the sweet dish, eg fruit salad in a bowl and gâteau on a plate.

SAVOURIES

A savoury, as the term indicates, is a savoury item served on varying shapes of toast, in tartlettes (round) and barquettes (boat-shaped) in bouchées, or flan rings or as an omelette or soufflé. The cutlery used is usually a side knife and sweet fork.

— *Savouries on toast* —

Anchovies	anchovy fillets laid side by side on toasted bread; finished with oil from the anchovies and decorated with sieved hard boiled yolk of egg and chopped parsley
Buck rarebit	a Welsh rarebit garnished with a poached egg
Haddock	flaked, poached haddock mixed with a cream sauce and spread onto the toast
Moelle	slices – approximately 6 mm ($\frac{1}{4}$ in) – of poached marrow dressed on hot buttered toast and coated with beurre noisette and decorated with chopped parsley
Mushrooms	shallow fried or grilled mushrooms, either sliced or whole, dressed on the toast and decorated with chopped parsley
Roes	shallow-fried roes dressed on toast, sprinkled with lemon juice and coated with beurre noisette
Sardines	sardines split with backbone removed, laid head to tail on toast and finished with oil from the sardines
Welsh rarebit	prepare ⅛ litre (¼ pt) béchamel (thin). Reduce ¼ litre (½ pt) beer to an ⅛ litre (¼ pt) and add to the béchamel. Stir in 113 g (4 oz) grated Cheddar. Season with salt, cayenne pepper and Worcester sauce. Bind with a liaison of one egg yolk. Pour on to fingers of toast and glaze under the grill.

— *Canapés* —

These are shaped pieces of bread, approximately 6 mm (¼ in) thick, brushed over with melted butter and placed under a salamander and coloured on both sides. Could be shallow fried.

Angels on horseback	poached oysters, wrapped in streaky bacon and grilled on skewers
Canapé Baron	garnish with slices of fried mushrooms, grilled bacon and poached bone marrow

Canapé Charlemagne	garnish with shrimps bound with a curry sauce
Canapé Nina	half small grilled tomato, garnished with mushroom head and a pickled walnut
Canapé Quo Vadis	grilled roes garnished with small mushroom heads
Canapé Ivanhoë	creamed smoked haddock, garnished with pickled walnut
Devils on horseback	stoned cooked prunes, stuffed with chutney and sprinkled with cayenne. Wrapped in streaky bacon and grilled on a skewer

— *Croûtes* —

These are shaped pieces of bread approximately 6 mm (¼ in) thick. Shallow fried.

Croûte Derby	spread with ham purée and garnished with a pickled walnut
Croûte Diane	partly cooked chicken livers (fried) wrapped in streaky bacon and grilled on a skewer
Croûte Windsor	spread with ham purée and garnished with small grilled mushrooms
Scotch Woodcock	scrambled egg garnished with a trellis of anchovies and studded with capers

Note: *Most of the above can be called either* canapés *or* croûtes

— *Tartlettes (round) or barquettes (boat-shaped)* —

These are normally made from unsweetened short crust pastry.

Charles V	soft roes mixed with butter and covered with a cheese soufflé mixture. Baked in the oven
Favorite	filled with a cheese soufflé mixture and slices of truffle. Garnished with slices of crayfish tails or prawns

Haddock	filled with diced haddock bound with a curry sauce. Sprinkled with breadcrumbs and gratinated

— *Bouchées* —

These are small puff pastry cases – a smaller edition of a vol-au-vent. They may be filled with various fillings such as shrimp, lobster, prawn and haddock.

Indienne	filled with curried shrimps and chutney

— *Flans* —

These are made from unsweetened short crust pastry.

Quiche Lorraine	flan made from unsweetened short crust pastry and filled with rashers of streaky bacon and slices of cheese. Covered with a savoury egg custard mixture and baked in the oven. Served hot

— *Omelettes* —

When served as a savoury, omelettes are usually made from two to three eggs and are flavoured in numerous ways such as parsley, anchovy, cheese, *fines herbes* (mixed herbs) etc.

— *Soufflés* —

These are made in soufflé dishes and can be produced using a variety of ingredients such as mushrooms, spinach, sardines, anchovies, haddock, cheeses etc.

DESSERT (fresh fruit and nuts)

Dessert is served on the lunch and dinner menu as an alternative to an entremet (sweet), whereas on a banquet menu it may appear as a course on its own. On an à la carte menu, dessert can be a section on its own.

Dessert may include all types of fresh fruits and nuts according to season, although the majority of the more popular items are available all the year round, due to the up-to-date means of transport between various parts of the world. Some of the more popular items are: dessert apples, pears, bananas, oranges, mandarines, tangerines, black and white grapes, pineapple and assorted nuts such as brazils etc. Sometimes candied fruits are included in the fruit basket.

BEVERAGES

Traditionally the term *beverages* on a menu referred to coffee but it has become more common for it to encompass tea, tisanes, milk drinks (eg hot chocolate) and other proprietary drinks such as Bovril and Horlicks.

Fairly rigid guidelines used to exist for the service of tea and coffee:

- Morning coffee was traditionally served in tea cups, with hot milk and white sugar only
- In the evening demi-tasse (half cups) were used for coffee, and cream might have been offered. Brown sugar was available as well as white
- Similarly with tea, breakfast cups were used in the morning and the smaller tea cup in the afternoon and in the evening – if you were lucky enough to be offered tea!
- Lemon was offered only with China tea and milk with other teas

Fortunately these so-called rules are not now so rigidly applied, meaning that more choice is now available. Both tea and coffee are more commonly available throughout the day with a choice of milks, creams (including non-dairy creamers), and sugars (including non-sugar sweeteners). The use of the demi-tasse for coffee is also on the decline.

— *Tea* —

Most of our tea comes from four main countries with each producing distinctive styles.

China   this is the oldest tea growing country and is best known for its delicately perfumed teas, examples of which are *Lapsang Souchong, Rose Pouchong, Orange Pekoe* and *Jasmine*

Ceylon (Sri Lanka)	these teas have a delicate light lemon flavour and are regarded as afternoon teas. They also lend themselves to being served iced
India	this country is the world's largest tea producer, the best known teas being *Dargeeling* and *Assam*
Kenya	produces medium flavoured teas

Other teas

Earl Grey	this is a blend of *Dargeeling* and *China* tea and is flavoured with bergamot
Tisanes	these are fruit flavoured and herbal teas examples of which include:

- **herbal teas:** camomile, peppermint, rosehip and mint
- **fruit flavoured teas:** cherry, lemon, blackcurrant and mandarin orange

Russian/lemon tea	this tea is generally made from Indian or Ceylon tea and served with slices of lemon, often in special tea glasses
Iced tea	this is a strong tea which has been cooled and then chilled in the refrigerator. It is served with lemon and sugar is offered

— *Coffee* —

Coffee beans are grown in many countries in the tropical and sub-tropical belt of South America, Africa and Asia. Brazil is the world's largest producer, with Columbia second.

The beans are roasted to varying degrees to bring out the coffee flavour, the most common degrees of roasting being:

- **Light/pale:** suitable for mild beans to preserve their delicate aroma
- **Medium:** gives a stronger flavour and well-defined character
- **Full:** popular in many Latin countries, often producing a bitterish taste

- **High:** this accentuates much of the strong and bitter aspects of the coffee and can destroy much of the original flavour

Most coffee beans are sold as named or branded blends containing a variety of coffees and roastings.

Methods of making coffee and grinds

Instant   This is made coffee which has then been dehydrated and which is reconstituted by adding boiling water.

Filter/drip   Uses a fine to medium grind coffee. This method involves pouring boiling water into a container which holds coffee and a filter paper or mesh. The coffee drips through to a lower container. This system may be used for a single cup of coffee or multiple cups. Machines are available and the method is widely used.

Jug   Uses a medium to coarse grind. The ground coffee is placed into a jug and boiling water added. After a time the grains will float to the surface. If a spoon is drawn over the surface, the grains will sink to the bottom, but it is still advisable to use a strainer when serving.

Cafetière   Uses medium grind. This is similar to the jug method but is made in a custom designed jug which has a plunger to act as a filter.

Vacuum infusion   Uses a medium grind. This method is characterized by the double pot or glass bowl and filter which many people know by the trade name 'Cona' – the company who make the glass equipment 'Cona'.

Espresso   Uses a fine grind. A process of forcing steam through a fine filter containing coffee. Often strong.

Cappuccino   Uses a fine grind. This is espresso coffee to which milk heated by steam has been added. Powdered chocolate is often sprinkled on the top of the frothing milk just prior to service.

Percolator   Uses a medium grind. A method of making coffee where the ground beans are contained in a filter which stands in the pot. This method has become less popular because the process boils the coffee as it is being made.

Turkish/Egyptian   Uses powdered coffee. This is made from dark roasted coffee in special copper pots. The pots are filled with water which is then boiled and sugar added. The powdered coffee is then added to produce a very strong beverage. Vanilla pods are sometimes added as flavouring.

Iced coffee   This is made from conventionally produced coffee which has been allowed to cool, often with the addition of milk. It is then chilled in the refrigerator. Whipped cream is often served on top.

Speciality coffees

These are coffees made in the conventional way to which spirits or liqueurs are added and then double cream floated on top. Sugar is necessary both for taste and to ensure that the cream will float. The main varieties of speciality coffee are:

Monks	Bénédictine
Russian	Vodka
Jamaican/Caribbean	Rum
Calypso	Tia Maria
Highland	Scotch whisky
Gaelic/Irish	Irish whiskey
Seville	Cointreau
Royale/Parisienne	Cognac

Note: *The names of these speciality coffees and the spirit they contain vary from establishment to establishment.*

3 Breakfast and afternoon tea

BREAKFAST

Breakfast is traditionally a British rather than Continental meal, originating from the days of the private house and family service where it was a substantial meal, consisting of six or seven courses. On the Continent breakfast was, and still is, much more of a snack as the midday meal is generally taken earlier and is more substantial. It has now become the trend in Britain for hotels to offer Continental breakfast inclusive in the room price with an additional charge for full English breakfast. In the market outside of hotels there has been a trend for a variety of sectors to offer some kind of breakfast service and it is not uncommon nowadays to find a variety of breakfast service in most major towns.

— English breakfast —

The full English breakfast consists of from two to eight courses. An example of a menu is shown opposite.
  Breakfast is usually taken in the order shown in the example, but any beverages are frequently placed on the table at the start of the meal.
  Bread items are also served with main courses.

— Continental breakfast —

Traditionally the Continental breakfast consisted of rolls and coffee. Nowadays it is not uncommon to find a Continental breakfast menu similar to that given opposite, excluding fish, eggs, meats and the cold buffet.

Breakfast Menu

Chilled Fruit Juices
Orange, Pineapple, Grapefruit, Tomato

Fresh and Stewed Fruits
Melon, Pineapple, Oranges, Apples
Prunes, Pears, Figs

Cereals
Porridge
Proprietary Brands of Breakfast Cereals

Fish
Finnan Haddock, Grilled Herring, Bloaters
Fried or Grilled Kippers, Fried Smelt
Fried or Grilled Plaice or Sole
Kedgeree

Eggs
Fried, Poached, Scrambled, Boiled
Plain or Savoury Omelette

Meat
Fried or Grilled Bacon
Fried or Grilled Sausages
Kidneys, Tomatoes, Mushrooms
Sauté Potatoes or Potato Cakes

Cold Buffet
York Ham, Calves' Tongue, Breakfast Sausage
Cheese

Breads
Toast, Rolls, Croissants, Brioches, Ryvita
Hovis and Procea
Danish Pastries

Preserves
Marmalade, Honey, Plum or Cherry Jam

Beverages
Tea, Coffee, Tisanes, Chocolate, Milk

AFTERNOON TEA

Afternoon tea may be classified into the two main types:
- **full afternoon tea**
- **high tea**

The main difference between the two is that *high tea* is a more substantial meal, including main course items such as grills, toasted snacks, fish and meat dishes, cold sweets and ices, in addition to the afternoon tea fare. The meat dishes frequently include pies and the fish dishes are usually fried. Vegetables and fried potatoes are included with some of the hot dishes.

An example of a *full afternoon tea* menu is given below. The items are usually served in the order listed, and beverages are served first.

Menu

Assorted Afternoon Tea Sandwiches:
Smoked Salmon, Sardine, Cucumber
Egg and Cress, Egg and Tomato
Gentleman's Relish

Hot Buttered Toast or Toasted Teacake
Crumpets

Brown and White Bread and Butter
Fruit Bread and Butter
Buttered Scones
Raspberry or Strawberry Jam

Gâteaux and Pastries

A cream tea *consists of two scones, jam (usually strawberry) and whipped or clotted cream, together with beverages.*

4 Foods in season

Name	Season	French
FISH (poisson)		
Barbel	June–March	Barbeau
Bream (sea)	July–end of December	Brème
Brill	August–March	Barbue
Cod	May–February best	Cabillaud
Dab	July–December	Limande
Eel	All the year (poor quality summer)	Anguille
Flounder	January–May	Flet
Haddock	All the year	Aiglefin
Hake	September to February	Merluche
Halibut	All the year	Flétan
Herring	Best September–April	Hareng
Lemon sole	October–March	Limande
Mackerel (red)	December–May	Rouget
Plaice	Best May–January	Plie/carrelet
Salmon	February–September	Saumon
Salmon trout	March–September	Truite saumonée
Smelt	October–May	Eperlan
Skate	October–May	Raie
Sole	All the year (poor quality spring)	Sole
Sturgeon	December–April	Esturgeon
Trout (river)	March–October	Truite de rivière
Turbot	February–September	Turbot
Whitebait	February–September	Blanchaille
Whiting	August–February best	Merlan

Name	Season	French
SHELLFISH (crustaces et mollusques)		
Crab	Preferably summer	Crabe
Crayfish	October–March	Ecrevisse
Crawfish	January–July	Langouste
Lobster	Preferably summer	Homard
Mussel	September–May	Moule
Oyster	1 September–30 April	Huître
Prawn	September–May	Crevette rose
Shrimp	All the year	Crevette grise
Scallop	September–April	Coquille St Jacques
BUTCHER'S MEAT (viande)		
Beef	All the year	Boeuf
Lamb	Best spring and summer	Agneau
Mutton	All the year	Mouton
Pork	September–end of April	Porc
Veal	All the year	Veau
GAME (feathered) (gibier)		
Wood grouse	12 August–12 December	Coq de bruyère
Partridge	1 September–1 February	Perdreau
Pheasant	1 October–1 February	Faisan
Ptarmigan	August–December	Ptarmigan
Quail	All the year	Caille
Snipe	August–1 March	Bécassine
Woodcock	August–1 March	Bécasse
Teal	Winter–spring	Sarcelle
Wild duck	September–March	Canard sauvage
Wood pigeon	1 August–15 March	Pigeon des bois

Name	Season	French
GAME (furred)		
Hare	1 August–end of February	Lièvre
Rabbit	Preferably autumn–spring	Lapin
Venison	Male best May–September	Venaison
	Female best September–January	
POULTRY (volaille)		
Chicken	All the year	Poulet
Duck	All the year	Canard
Duckling	April–May–June	Caneton
Goose	Autumn–winter	Oie
Gosling	September	Oison
Guinea fowl	All the year	Pintade
Spring chicken	Spring (cheapest)	Poussin
Turkey	All the year	Dinde
FRUIT (fruits)		
Apple	All the year	Pomme
Apricot	May–September	Abricot
Blackberry	Autumn	Mûre
Cherry	May–July	Cerise
Cranberry	November–January	Airelle rouge
Currant (black & red)	Summer	Groseille
Damson	September–October	Prune de damas
Gooseberry	Summer	Groseille à maquereau
Greengage	August–September	Reine-Claude
Grapes	All the year	Raisin
Melon (cantaloup)	May–October	Melon
Nectarine	June–September	Brugnon

Name	Season	French
Peach	All the year (best June–September)	Pêche
Pear	All the year (best autumn–winter)	Poire
Plum	July–October	Prune
Pineapple	All the year	Ananas
Raspberry	Summer	Framboise
Rhubarb	January–July	Rhubarbe
Strawberry	June–September	Fraise

VEGETABLES (légumes)

Name	Season	French
Artichoke globe	Best summer–autumn	Artichaut
Artichoke Jerusalem	October–March	Topinambour
Asparagus	May–July	Asperge
Beetroot	All the year	Betterave
Broad bean	July–August	Fève
Broccoli	October–April	Brocolis
Brussels sprout	October–March	Chou de Bruxelles
Cabbage	All the year	Chou
Cauliflower	All the year	Choufleur
Carrot	All the year	Carotte
Celery	August–March	Céleri
Celeriac	November–February	Céleri-rave
Cucumber	Best summer	Concombre
Chicory (Belgian)	Best winter	Endive belge
Egg plant	Best summer–autumn	Aubergine
French bean	July–September	Haricot vert
Leek	October–March	Poireau
Lettuce	Best summer	Laitue
Mushroom	All the year	Champignon
Onion	All the year	Oignon
Pea	June–September	Petit pois
Parsnip	October–March	Panais
Radish	Best summer	Radis
Runner bean	July–October	Haricot d'espagne
Salsify	October–February	Salsifis

Name	Season	French
Sea kale	January–March	Chou de mer
Shallot	September–February	Echalotte
Spinach	All the year	Epinards
Swede	December–March	Rutabaga
Sweetcorn	Autumn	Maïs
Tomato	All the year (best summer)	Tomate
Turnip	October–March	Navet
Vegetable marrow	July–October	Courgette

FRESH HERBS (fine herbes)

Bay-leaf	September	Laurier
Borage	March	Bourrache
Chervil	Spring–summer	Cerfeuille
Fennel	March	Fenouil
Garlic	All the year	Ail
Garlic (clove)	All the year	Gousse d'ail
Marjoram	March	Marjolaine
Mint	Spring–summer	Menthe
Parsley	All the year	Persil
Rosemary	August	Romarin
Sage	April–May	Sauge
Thyme	September–November	Thym
Tarragon	January–February	Estragon

5 Glossary of cuisine terms

À bleu	rare, as in the cooking of steak
À l'Anglaise	in English style, i.e. plainly roasted, boiled, fried with simple accompaniments
À la broche	cooked on a skewer
À point	medium cooked
Aromatic	any herb, plant or root which gives off an agreeable smell
Au four	cooked in the oven
Au naturel	uncooked
Baba	yeast sponge or bun
Bain marie	hot water bath or well
Bard (to)	cover the breast of a roasting bird with a slice of fat bacon
Baron	saddle and two legs of a cow or sheep, roasted in one piece
Baste (to)	pour fat over a piece of meat whilst it is roasting
Bat (to)	flatten fillets or cutlets
Beurre manié	mixture of butter and flour used to thicken sauces
Bien cuit	cooked well done
Bind (to)	thicken a sauce
Bisque	a fish soup made with shellfish
Blanche (to)	partially cook something by bringing it to the boil and allowing it to cool
Blanquette	white stew with final sauce poured over
Blind (to cook)	bake a pastry case separately from its filling
Bouquet garni	bay leaf, sprig of thyme and a stalk or two of parsley

Braise (to)	cook something in a small amount of liquid in a braising dish or covered casserole
Brioche	type of yeast roll
Brown (to)	colour vegetables or meat in a frying pan or oven
Brunoise	vegetables chopped up into cubes approx 1–3 mm square
Bouillon	fluid obtained by boiling meat or beef bones in seasoned water
Canapé	small piece of fried bread or pastry enclosing some form of savoury mixture
Caramel	sugar heated until it forms a brown fluid
Carcase	bony structure of animal or bird (especially poultry)
Carte du jour	menu (literally *card*) of the day
Chaud	hot
Chiffonnade	sorrel, lettuce or chervil chopped and mixed with melted butter
Clarify (to)	(1) make clear a bouillon which is cloudy (2) melt butter in a double boiler
Cocotte	small earthenware container
Concassé	rough chopped (skinned and seeded tomato)
Confiture	jam
Coquille, en	cooked in a shell, usually a scollop shell
Court-bouillon	water with wine or vinegar and aromatics added, in which fish is cooked
Couvert	cover or place setting
Croûte	piece of crisply fried bread
Croûte, en	cooked in pastry
Croûton	small piece of fried bread normally used to accompany soups
Darne	thick slice of a round fish including the central bone
Dariole	small beaker-shaped mould
Daube (to cook in)	similar method of cooking to braising
De-bone (to)	remove the bone from a piece of meat
De-glaze (to)	remove juices which have adhered to sides of a pan or casserole in which food has been browned
De-grease (to)	remove grease or fat from a liquid

Demi-tasse	literally half cup, small coffee cup
Dorer	to egg wash pastry, or milk wash
Double saucepan	form of saucepan in which water in a lower pan heats the contents of an upper pan
Drain (to)	allow the liquid to run out of a substance
Dress (to)	apply garnishing to a dish
Duxelle	form of mashed mushroom
Emancer	sliced
Emincé	minced
Entrée	(1) the first meat dish served
	(2) on a formal menu, the dish served before the roast
Escalopes	small, flattened slices of meat or fish
Essence	reduction of some substance such as tomato, shrimp, etc
Faggot of herbs	bunch of herbs for flavouring, generally known by the French term, *bouquet garni*
Farce	stuffing
Fin (to)	remove the fins or barbs from a fish
Fines herbes	mixed herbs
Flambé (to)	burn a spirit or fortified wine over the food
Fleurons	small crescent shaped pieces of puff pastry used as garnish
Frappé	chilled
Fricassée	white stew
Froid	cold
Fumé	smoked
Glacé	glazed or iced
Glaze (to)	(1) coat meat with a hard surface by pouring over its gravy and then heating for a short period
	(2) coat cold meat with meat glaze
Guéridon	side table for service; may also be a trolley
Haché	minced
Julienne	fine strips of vegetables or cooked meats
Jus	natural cooking liquor
Kneaded butter	the same as *beurre manié*
Lard (to)	insert pieces of fat bacon into a joint or bird
Lardons	strips of fat bacon used for larding meat or fish

Line (to)	cover the interior of a mould with a thin coating of, for example, aspic or ice-cream, before filling it with some other mixture
Macédoine	diced vegetables or fruit
Marinate (to)	soak meat or game in wine or oil to flavour it
Mince (to)	chop up something very finely
Mirepoix	roughly chopped vegetables
Nap (to)	cover food with a thick sauce
Noisette butter	butter cooked until it is a light brown colour
Panada	basis of a soufflé
Paupiettes	stuffed and rolled slices of meat or fish
Paysanne	thin slices of vegetables
Plat du jour	special dish of the day
Poach (to)	cook food in a liquid that is hardly simmering
Poêlé	pot roast
Porringer	see *double saucepan*
Pound (to)	crush food with a heavy utensil
Quenelles	Kind of dumpling made from various kinds of meat or fish and made into various shapes
Reduce	to boil down
Refresh (to)	dip hot food into cold water to stop it cooking further
Rissole (to)	brown slowly in fat
Roux	basis of flour sauces
Saignant	cooked underdone (medium rare)
Sauté (to)	fry in a small amount of very hot fat
Scald (to)	immerse something, briefly, in boiling water
Skim (to)	remove the top surface from a liquid
Steak	thick slice of meat or fish
Stew (to)	cook by simmering in a closed pan with little liquid
Stiffen (to)	make thicker
Strain (to)	remove the liquid from a substance, usually in a strainer
Suprême	cut chicken comprising the breast and wing
Tasse, en	in cup
Timbale	similar to bain-marie but using ice for cold food
Tronçon	portion of fish cut across the body
Truss (to)	tie up a fowl or meat with string

Voiture   trolley, for example for hors d'oeuvre or
            sweet
Wagon   another name for voiture

SECTION

On wine and other drinks

1   Introduction to wine

WINE MAKING

Wine is as old as history. The spread of the vine started with the Assyrians and was carried on by the Phoenicians, the Greeks, Romans and religious orders. Latterly it was continued by exiles such as the Italians, Germans, Spanish and French who established vineyards wherever they formed into large communities in lands compatible with the cultivation of the vine.

Wine is fermented grape juice, the best quality being produced mostly in the latitudes 30–50 ° north and south of the Equator. There are four main families of wine-producing vines: *Vitis vinifera* (wine bearing) which produces all the great wines, and *Vitis riparia*, *Vitis labrusca* and *Vitis rupestris* which make less fine wines.

Wine making has never been easy and wine growers down the ages have had to contend with and protect their vines from all kinds of pests and diseases such as *oidium*, *mildew*, *grey rot*, *cochylis*, *rougeot*,

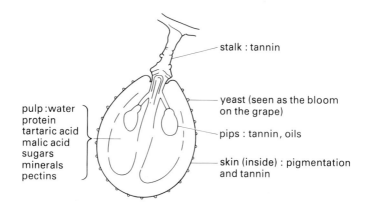

stalk : tannin

yeast (seen as the bloom on the grape)

pips : tannin, oils

skin (inside) : pigmentation and tannin

pulp : water
protein
tartaric acid
malic acid
sugars
minerals
pectins

chlorose, *coulore* and *browning*. The worst pest of all was the dreadful *Phylloxera vastatrix*. This aphid, which lives in a gall on the underside of the vine leaf, or on the roots, was brought from North America to Europe in the latter half of the nineteenth century. It attacked the vine roots and devastated most of the world's vineyards. Later on it was discovered that American rootstocks (eg *Vitis labrusca*) were resistant to the disease. Thus, today most vines are grafted on to American stocks.

— *The grape* —

The grape, which has developed on the vine for about 100 days after flowering, is made up of *skin*, *stalk*, *pips* and *pulp*.

Skin

The outer skin (cuticle) is coated with a waxy film (bloom) on which there are, by the time the grape is ripe, about 100,000 wine yeasts plus 10 million wild yeasts and other micro-organisms. Wine yeasts

(*Saccharomyces ellipsoideus*) spend winter in the intestines of animals and in spring they are disseminated to alight on flowers and plants. In summer the wind and insects carry them to settle on the ripening grape skins where they are trapped by the waxy film and form a downy coating on the grape. Each yeast contains thousands of minute enzymes (ferments) and it is these, when they come in contact with the sugars in the grape juice and air, that causes a chemical reaction, known as *fermentation*, to take place.

The inside of the grape skin contains colouring matter which is extracted during fermentation by the alcohol.

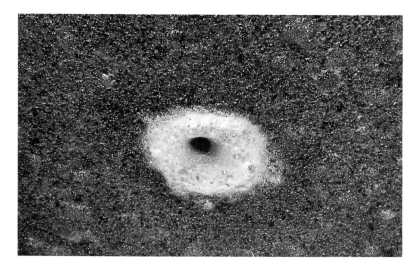

Stalks

When used in winemaking (red wine in particular), the stalks provide *tannic acid* which gives body and keeping qualities to wine. If over used, this acid can cause wine to become too astringent and nasty. However, when used correctly it gives wine a dry flavour and a good grip on the palate. During the maturing process of wine, tannic acid helps to coagulate the fining agent as the wine is being clarified.

Pips

The pips, if crushed, impart tannic acid and oils to the wine.

Pulp

The pulp provides the grape juice or *must* and contains:

78–80%	water
10–25%	sugar
5–6%	acids

Water: makes up the bulk of wine.

Sugar: which was formed in the grapes by sunlight is of two kinds, *grape sugar* (dextrose and glucose) and *fruit juice* (levulose and fructose). It is found in about equal quantities in the *must* and is essential to the fermenting process.

Acids: tartaric, malic, tannic and citric help to give keeping quality, freshness, brilliance and balance. When they come in contact with alcohol, *esters* are formed and it is these that give a wine its bouquet.

> *The* must *(unfermented grape juice) will also have trace elements of nitrogenous compounds such as albumen, peptones, amides, ammonium salts and nitrates, as well as potassium, phosphoric acid and calcium, all of which have an influence on the eventual taste of the wine.*

— How wine is made —

The making of wine starts with the gathering of the grapes (the *vintage*). The grapes are crushed to produce the must. This is run off into huge temperature controlled vats where fermentation takes place. Fermentation, which can last from a week to three weeks, is the action of yeast (found on the bloom covering the outside skin of the grape) on the sugar in the must, converting it into ethyl alcohol and carbon

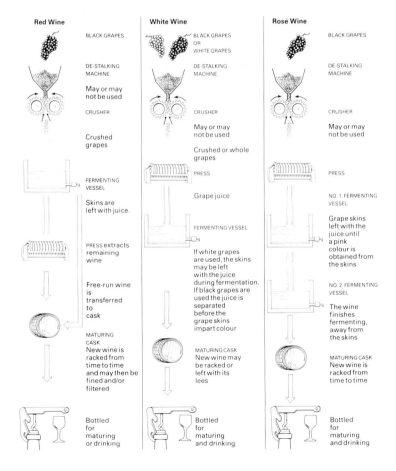

Red Wine

BLACK GRAPES

DE-STALKING MACHINE
May or may not be used

CRUSHER

Crushed grapes

FERMENTING VESSEL
Skins are left with juice.

PRESS extracts remaining wine

Free-run wine is transferred to cask

MATURING CASK
New wine is racked from time to time and may then be fined and/or filtered

Bottled for maturing or drinking

White Wine

BLACK GRAPES OR WHITE GRAPES

DE-STALKING MACHINE

CRUSHER
May or may not be used

Crushed or whole grapes

PRESS

Grape juice

FERMENTING VESSEL

If white grapes are used, the skins may be left with the juice during fermentation. If black grapes are used the juice is separated before the grape skins impart colour

MATURING CASK
New wine may be racked or left with its lees

Bottled for maturing and drinking

Rosé Wine

BLACK GRAPES

DE-STALKING MACHINE

CRUSHER
May or may not be used

PRESS

NO. 1. FERMENTING VESSEL
Grape skins left with the juice until a pink colour is obtained from the skins

NO. 2. FERMENTING VESSEL
The wine finishes fermenting, away from the skins

MATURING CASK
New wine is racked from time to time

Bottled for maturing and drinking

dioxide gas (CO_2) through a series of complex biochemical reactions. The gas forms bubbles on the surface before escaping into the air.

There are two types of yeast: *wild yeasts* which start the fermentation and are killed when the alcohol level reaches 4% by volume and *wine yeasts* which then take over and can continue to convert the sugar up to a maximum of 16% alcohol by volume. Most table wines however, have an alcohol strength of between 10% and 14%.

The new wine is usually transferred to casks where is will be racked from time to time as it matures. The purpose of racking is to eliminate the *lees* (sediment or deposit) in the wine. This deposit is left behind as the wine is moved to a fresh cask and so, with each racking, the wine becomes clearer. Before being bottled the wine is *fined* to get rid of unwanted particles held in suspension. Isinglass, egg whites, gelatine and dried albumen are particularly good fining agents as they attract the unwanted particles and drag them down to the bottom of the cask, thus leaving behind brilliant wine which may or may not be filtered before bottling. The longer the wine matures in cask the less time it needs in bottle. Great wines are matured for many years in bottle.

— Factors that influence the quality of wine —

1 Climate and microclimate
2 Nature of the soil and subsoil
3 Vine family and grape species
4 Method of cultivation – *viticulture*
5 Chemical composition of the grape
6 Yeast and fermentation
7 Methods of wine making – *vinification*
8 Luck of the year
9 Ageing and maturing process
10 Method of shipping or transportation
11 Storage temperature

— *The top ten wine producing countries by volume* —

1 Italy	6 USSR
2 France	7 Portugal
3 Spain	8 West Germany
4 Argentina	9 Romania
5 USA	10 South Africa

This order is according to the volume of wine produced and it has been known to change periodically.

WINE STYLES

— *Red wines* —

Red wine comes from black grapes that are fermented throughout in the presence of their skins and sometimes stalks. As fermentation continues the alcohol generated draws colour from the inside of the skins.

— *Rosé wine* —

Rosé wine comes from black grapes without the stalks and is made in a similar way to red wine, but the juice is separated once the desired degree of pinkness has been achieved. Fermentation is completed in a separate vessel. Alternative methods are either to press the grapes so that some colour is extracted from the skins or to blend red and white wine together.

— *Blush wine* —

Blush wine is a new style of rosé wine originating in California in which the black grape skins are left to macerate for only a very short period with the must. The resulting wine has a blue-pink hue with copper highlights.

— *White wine* —

White wine comes from either black or white grapes. When black grapes are used, the juice is quickly moved to another vessel to begin and complete its fermentation. All grape juice is colourless initially. The juice from white grapes is usually left with the grape skins until fermentation is completed.

— *Sparkling wine* —

When making sparkling wine, a sugar solution and special yeast culture are firstly added to dry table wine. The wine is then sealed and a secondary fermentation is allowed to take place:

(a) in a bottle (*méthode Champenoise* or *méthode traditionelle*)
(b) in a tank (*méthode cuve close* – also known as the Charmat or bulk method)
(c) in a bottle and then the wine is transferred under pressure to a tank or vat where it is filtered and rebottled (*méthode transvasage* – transfer method).

Sparkling wine can also be made by injecting CO_2 into the chilled vats of still wine and then bottling the wine under pressure (*méthode gazefié*).

— *Organic wines* —

These wines, also known as 'green' or 'environmentally friendly' wines, are made from grapes grown without the aid of artificial insecticides, pesticides or fertilizers. The wine itself will not be adulterated in any way, save for minimal amounts of the traditional preservative, sulphur dioxide, which is controlled at source.

— *Alcohol-free, de-alcoholized wines and low alcohol wines* —

Alcohol-free	maximum	0.05% alcohol
De-alcoholized	maximum	0.50% alcohol
Low alcohol	maximum	1.25% alcohol

These wines, leaving aside the commercial angle, have been produced as a response to today's emphasis on health and fitness. Specifically, they are aimed at the designated driver, the weight watcher, the medical patient, the sports enthusiast, the business man, those on diets with a restricted alcohol intake and even at the non-drinker. The wines are made in the normal way and the alcohol is then removed either by (a) the hot treatment, *distillation*, which unfortunately removes most of the flavour as well, or, more satisfactorily, by (b) the *cold filtration process*, also known as *reverse osmosis*. This removes the alcohol by mechanically separating or filtering out the molecules of alcohol through membranes made of cellulose of acetate. The wine is repeatedly passed through the membranes which filter out the alcohol and water and leaves behind a syrupy wine concentrate. To this, at a later stage, water and a little *must* are added, thus preserving much of the flavour or mouthfeel of the original wine.

— *Vins doux naturels* —

These are sweet wines that have had their fermentation muted by the addition of alcohol in order to retain their natural sweetness. Muting takes place when the alcohol level reaches between 5% and 8% by volume. They have a final alcoholic strength of about 17% by volume.

— *Fortified wines* —

Fortified wines such as sherry, port and Madeira are those that have been strengthened by the addition of alcohol, usually a grape spirit.

— *Aromatized wines* —

These are flavoured and usually fortified. Typical examples are vermouths and Commandaria.

GRAPES FOR WINE

The following is a list of the principal grapes used in wine production in the main wine growing regions of the world.

— *White grapes* —

Aligoté	Burgundy
Bacchus	Mainly Germany but also grown in England
Blanc Fumé	see *Sauvignon Blanc*
Blanquette	see *Columbard*
Bual (Boal)	Madeira
Catawba	North American States (*Vitis labrusca*)
Chardonnay	Champagne, Burgundy, California, Eastern Europe, South America, Australia and New Zealand; sometimes referred to as *Pinot Chardonnay*
Chasselas	France – mainly the Loire (eg Pouilly-sur-Loire); Central Europe; known as *Gutedel* in Germany and *Fendant* in Switzerland
Chauche Gris	see *Grey Riesling*
Chenin Blanc	Loire (eg Vouvray, Saumur, Côteaux du Layon), California, South America, South Africa (known as *Steen*), Australia
Clairette	Mainly southern France
Colombard	France, especially in Cognac (also known as *Blanquette* elsewhere in France), California (known as *French Colombard*)
Delaware	North American States (*Vitis labrusca*)
Fendant	see *Chasselas*
Folle Blanche	France, especially Cognac, Armagnac and the Loire, and California; also known as *Picpoule*
French Colombard	see *Colombard*

113

Furmint	Hungary (eg Tokay); also known as *Sipon* in Hungary
Gewürztraminer	France – in Alsace (also in the Jura where it is known as *Savignin*); Germany, Austria, Australia, northern Italy, California
Grey Riesling	France, California; real name *Chauche Gris*
Gutedel	see *Chasselas*
Johannisberg Riesling	Germany – mainly Rhine and Mosel; France – in Alsace and the Jura; Central Europe, Australia, California; also called *White Riesling*
Listan Palomino	see *Palomino*
Malmsey	see *Malvasia*
Malvasia	California, Mediterranean; known as *Malmsey* in Madeira
Malvoisie	see *Pinot Gris*
Melon de Bourgogne	France – in Muscadet; also known as *Muscadet*
Müller-Thurgau	England, Germany, Austria, Central Europe
Muscadelle	Mainly in Australia, South Africa and some in France (in Bordeaux)
Muscadet	see *Melon de Bourgogne*
Muscat	France (eg Muscat de Beaumes-de-Venise), California, Spain, Italy (eg Asti Spumante), Mediterranean
Palomino	Spain for sherry; also known as *Listan Palomino*
Pedro Ximénez	Australia, California, South Africa, Spain for sherry
Picpoule	see *Folle Blanche*
Pinot Blanc	Burgundy, Alsace, Germany (where it is known as *Weissburgunder*), Italy, California
Pinot Chardonnay	see *Chardonnay*
Pinot Grigio	See *Pinot Gris*
Pinot Gris	France – mainly Alsace; Germany, Switzerland, northern Italy; known as *Pinot Grigio* in Italy, *Ruländer* in Germany, *Tokay* in Alsace and *Malvoisie* in Switzerland
Riesling	see *Johannisberg Riesling* and *Sémillion*
Ruländer	see *Pinot Gris*

114

Saint Emilion	see *Ugni Blanc*
Savignin	see *Johannisberg Riesling* and *Gewürztraminer*
Sauvignon Blanc	France – mainly Loire Valley (eg Sancerre and Pouilly Fumé) and Bordeaux (eg Graves and Sauternes); Chile, Australia, California; also called *Blanc Fumé*
Sémillon	France – mainly Bordeaux (eg Graves and Sauternes); South America, South Africa, Australia, California; often called *Riesling* in Australia
Sercial	Madeira
Seyval Blanc	North American States, Canada, France, England
Silvaner	see *Sylvaner*
Sipon	see *Furmint*
Steen	see *Chenin Blanc*

Sylvaner	Central Europe – mainly Germany; France – in Alsace; California; also known as *Silvaner*
Tokay	see *Pinot Gris*
Trebbiano	Italy (eg Soave, Orvieto, Frascati), France, California; also known as *Ugni Blanc*
Ugni Blanc	France mainly Cognac and Armagnac where it is known as *Saint Émilion*; also see *Trebbiano*

Verdelho	Madeira
Verdicchio	Mainly central Italy
Viognier	France mainly Rhône (eg Condrieu)
Welsch Riesling	Europe; no relation to *Johannisberg Riesling*
Weissburgunder	see *Pinot Blanc*
White Riesling	see *Johannisberg Riesling*

— *Red grapes* —

Baco Noir	North American States, France
Barbera	Italy – in Piedmont; South America, California
Breton	see *Cabernet Franc*
Brunello	Italy – in Tuscany (eg Brunello di Montalcino)
Cabernet Franc	France – mainly Loire (eg Cabernet d'Anjou, Chinon) and Bordeaux; Italy, California; also known as *Breton* in France
Cabernet Sauvignon	France – mainly Bordeaux and Provence; Chile, Bulgaria, California, Spain, Australia, almost everywhere
Carignan	France – mainly Rhône and Provence; California, Spain, North Africa
Chancellor Noir	North American States, southern France
Charbono	California
Cinsault	France
Concorde	North American States, California (*Vitis labrusca*)
de Chaunac	North American States, Canada
Dolcetto	Italy – in Piedmont
Duriff	see *Petit Sirah*
Gamay	France – mainly Beaujolais but also Loire; Switzerland, California
Gamay Beaujolais	California
Grenache	France – mainly south and southern Rhône (eg Châteauneuf-du-Pape, with other grapes, and for Tavel Rosé and Lirac), California, Spain
Lambrusco	Italy – in Emillia-Romagna

Malbec	France – in Bordeaux; Cahors, Argentina, California
Merlot	France – mainly Bordeaux; north Italy, Switzerland, California, South America, South Africa
Meunier	France – in Champagne; also called *Pinot Meunier*
Nebbiolo	Italy – in Piedmont and Lombardy (eg Barolo, Barbaresco); California
Petit Sirah	California; also called *Duriff*
Pinot Meunier	see *Meunier*
Pinot Noir	France – mainly Champagne and Burgundy, but also in the Loire; England, Switzerland, Germany (known as *Spätburgunder*), Eastern Europe, California, South America

Ruby Cabernet	California
Sangiovese	Italy – in Tuscany and Emillia-Romagna (eg for Chianti, but blended with other grapes)
Shiraz	see *Syrah*
Spätburgunder	see *Pinot Noir*
Syrah	France – mainly northern Rhône (eg Hermitage, St Joseph, Cornas, also used in blend for Châteauneuf-du-Pape); Australia; also known as *Shiraz*
Tempranillo	Spain – in Rioja; Argentina
Zinfandel	California; also used for blush and rosé wines

FAULTS IN WINE

Faults occasionally develop in wine as it matures in bottle. Nowadays, through improved technique and attention to detail regarding bottling and storage, faulty wine is a rarity. However, occasionally you may be unfortunate enough to come across a rogue bottle. Here are the more common causes.

Corked wines

These are wines affected by a diseased cork caused through bacterial action or excessive bottle age. The wine tastes and smells foul. Not to be confused with *cork residue* in wine which is harmless.

Maderization or oxidation

Caused by bad storage: too much exposure to air, often because the cork has dried out in these conditions. The colour of the wine browns or darkens and the taste very slightly resembles Madeira, hence the name. The wine tastes 'spoilt'.

Acetification

Caused when the wine is overexposed to air. The vinegar microbe develops a film on the surface of the wine and acetic acid is produced making the wine taste sour resembling wine vinegar (*vin aigre*).

Tartrate flake

This is the crystallization of potassium bitartrate. These crystal-like flakes, sometimes seen in white wine, may cause anxiety to some customers as they spoil the appearance of the wine which is otherwise perfect to drink. If the wine is stabilized before bottling, this condition should not occur.

Excess sulphur dioxide (SO_2)

Sulphur dioxide is added to wine to preserve it and keep it healthy. Once the bottle is opened the stink will disappear and, after a few minutes, the wine is perfectly drinkable.

Secondary fermentation

This happens when traces of sugar and yeast are left in the wine in bottle. It leaves the wine with an unpleasant, prickly taste that should not be confused with the *pétillant*, *spritzig* characteristics associated with other styles of healthy and refreshing wines.

Foreign contamination

Examples include splintered or powdered glass caused by faulty bottling machinery or re-used bottles which previously held some kind of disinfectant.

Hydrogen sulphide (H_2S)

The wine has the characteristic smell of rotten eggs. Throw it away.

Sediment, lees, crust or dregs

Organic matter discarded by the wine as it matures in cask or bottle. It can be removed by racking, fining or, in the case of bottled wine, by decanting.

Cloudiness

This is caused by suspended matter in the wine disguising its true colour. It may be due to extremes in storage temperatures.

THE WINE LIST

Wine lists, be they grandiose, printed and illustrated or a typed sheet in a folder, can in fact be very intimidating to the newcomer to wine. It is best to remember that they are merely a list of drinks held for sale in an establishment. Lists come in two styles: the *restaurant list*, which shows the complete range of beverages on offer, and the *banquet list*, which shows a more limited selection of the more commercial and popular items.

The pleasure of eating out is being enjoyed increasingly by more and more people, and the young especially are great experimenters in their choice of food and its subsequent marriage to wine. However, some people tend to shy away from wines with difficult to pronounce names, but a perceptive wine waiter will quickly gauge the situation, offering advice without causing embarassment. It is often better to order wine by the *bin number* – each wine listed is given a certain number corresponding to the number under which the wine is stored in the cellar. So you simply order a bottle of number 10 or whatever.

Wines featured on the wine list should complement the menu both in style and price and should encourage the more adventurous to experiment with wines that may not be readily available in other commercial outlets. Rarely, however, except perhaps in the more de luxe restaurants, will customers purchase wine to accompany a meal when the price of the wine exceeds the price of the meal itself. Fair mark-ups which come proportionately smaller as the wine gets more expensive will encourage people to experiment and buy good wine.

A wine list should also take into account those who dine alone or the couple with different preferences – one who likes white wine and the other red. Good wine, other than house wine, should be available by the glass and there should be some choice in half bottles. People on diets, organic wine lovers, those with illnesses such as diabetes, the ever increasing number who strive to lead healthy lifestyles and, of course, car drivers should all be catered for, and so low or de-alcoholized wines are a must nowadays on every drinks list.

A wine list does not have to be big to be good, but it should show a well balanced selection of wines not only in terms of the country of origin but also of the grape used. Above all, it should be informative, giving the bin number, the name of the wine (sometimes with descriptive comments if appropriate), the vintage year if applicable, the producer's, bottler's or shipper's name, whether the wine is bottled at source (ie estate or château bottled) and the price per magnum, bottle or half bottle.

The sequence in which the drinks are listed is traditional. First come the *apéritifs*, then the *wines* and other drinks and finally the *digestifs*. Wine lists can also be presented according to price bands which some people find particularly helpful in making their choice.

A wine list is a signpost for profitable sales. It makes a statement about the establishment, presenting the character and image the owners want to project.

2 Care and service of wine

SERVICE OF WINE

Whether at home or in a restaurant, one of the main concerns of any host is to have the wine to be offered to the guests properly temperatured. This is a basic requirement and gives the wine a real chance to shine. Ideally, one should know the food on offer so that a wine can be organized in advance and the appropriate temperature assured.

— Opening techniques —

Firstly cut away the top of the capsule to expose the cork. Above all, use a corkscrew with a wide thread so that the cork can be levered out without any crumbling. The cork should leave the bottle with a sigh as if sad to depart after such long contact. Smell the cork which should smell of wine. Sometimes a wine may have a bit of bottle stink due to stale air being lodged between cork and wine, but this soon disappears as the wine is exposed to air.

Red wine bottles are usually placed on a coaster beside the host, whereas sparkling, white, blush and rosé bottles should be placed in a wine bucket holding ice and water up to the neck of the bottle. In a restaurant it is very important to show the host the bottle with the label uppermost and to nominate the wine verbally so there is no confusion regarding the wine or vintage ordered.

— Pour, twist and take —

When pouring wine, the neck of the bottle should be over the glass, but not resting on the rim in case of an accident. Care should be taken to avoid splashing and, having finished pouring, the bottle should be twisted as it is being taken away. This will prevent drips of wine falling on the tablecloth or on someone's clothes. Any drips on the rim of the bottle should be taken away with a clean service cloth or napkin.

The host gets a little taster and decides that the wine is perfectly sound for drinking. Service proceeds on the right from the right around the table with the label clearly visible and with the host's glass being finally topped-up to the customary two-thirds full. Later, if another bottle of the same wine is ordered, the host should be given a fresh glass from which to taste the new wine.

— Opening and serving Champagne and sparkling wines —

Champagne and sparkling wines can be a bit of a problem. The bottle should not be shaken on its journey to the table and the wine must be well chilled. This helps control the effervescence and imparts the refreshing qualities associated with such wines.

Loosen or take off the wire muzzle. Holding the bottle at an angle of 45 ° in one hand and the cork in the other, twist the bottle *only* and, when the cork begins to move, restrain it by pushing it almost back into the neck. Soon the cork will leave the bottle quietly (never with a loud pop or bang). Should the cork prove stubborn and reluctant to leave the bottle, soak a napkin in hot water, wrap it around the neck of the bottle and movement will quickly occur.

Be extra careful when opening sparkling wines. Always keep the palm of your hand over the cork to prevent accidents to the eyes and elsewhere. Hold the bottle with your thumb in the punt and pour the wine against the inside of the glass – there should be a nice mousse but no frothing over.

— Decanting —

The main reasons for decanting an old red wine are (*a*) to separate it from the sediment, (*b*) to allow the wine to breathe and (*c*) to develop its bouquet. Fine old red wines and some ports which have spent most of their lives maturing in bottle throw a deposit or crust which, if allowed to enter the glass, would sully the appearance of the wine. This deposit forms as the wine ages and consists of tannins, bitartrates of

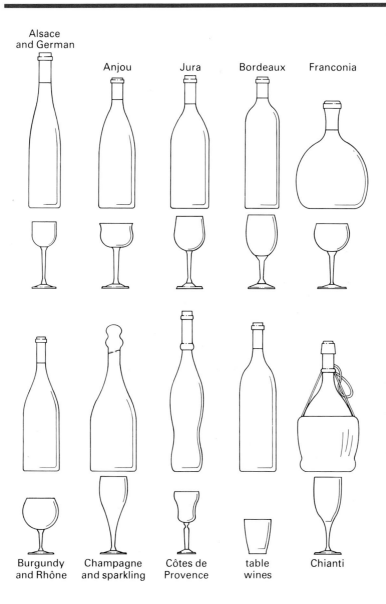

Alsace and German

Anjou

Jura

Bordeaux

Franconia

Burgundy and Rhône

Champagne and sparkling

Côtes de Provence

table wines

Chianti

calcium and magnesium and colouring matter. It makes the wine cloudy and can cause it to taste of lees.

Decanting is the movement of wine from its original container to a fresh glass receptacle, leaving the gunge or sediment behind. It is best to stand the bottle upright for two days before decanting to give the sediment a chance to settle at the bottom.

- Extract the cork carefully – it may disintegrate because of long contact with alcohol, so be wary
- Place a light behind the shoulder of the bottle, a candle if you are decanting in front of guests, but a torch, light bulb or any light source at home or in the cellar will do
- Carefully pour the wine through a funnel into an absolutely clean decanter. The light will reveal the first sign of sediment (known as *beeswing* in port) entering the neck of the bottle
- At this stage stop pouring into the decanter but continue pouring into a glass which should be handy. The latter wine when it settles can be used as a taster or for sauces in the kitchen and can provide a nice talking point for guests.

The quick method for decanting is to place a coffee filter or perfectly clean muslin in a funnel which should be in the neck of the decanter as the wine flows in. The vogue nowadays is also to decant younger red wines, simply because exposure to air improves the bouquet, and softens and mellows the wine. Of course, the host's permission must always be given before decanting a wine in a restaurant. Decanting also enhances the appearance of the wine, especially when presented in a fine wine decanter.

At home, experiment by pouring young and robust red wine into glasses 15 minutes before the meal – not only will the wine taste smoother, but the room will be pleasingly redolent of wine as the guests enter.

Very old red wine breaks up with too much exposure to air. It is best to stand such a bottle for a few days to allow the sediment to settle in the bottom. Then open the bottle just before the meal is served and pour the wine very carefully straight into the glass with the bottle held in the pouring position as each glass is approached. This prevents the wine slopping back to disturb the sediment. Sufficient glasses should be available to finish the bottle, thereby ensuring that the wine does not re-mingle with its sediment at the end of service.

Wine cradles or wine baskets are useful when taking old red wines from the cellar as they hold the bottle in the binned position, thus leaving the sediment undisturbed. If great care is taken when pouring, you can avoid disturbing the sediment by using the cradle, but the skill requires expertise and a large hand to span the cradle. It has become fashionable to serve red wines in a wine basket or cradle. Whilst this practice appeals to some, there is no technical argument for doing so.

Carafe wines, also known as house wines, should be pleasant to drink and may be red, white or rosé. Just as the quality of soup tells you about the chef's approach to cooking, so the house wine reveals the restaurant's attitude to wine. Unlike the decanter, which is often made of cut glass and has a stopper fitted, a carafe is usually plain, of clear glass and without a stopper.

— *Recommended service temperatures* —

Champagne and sparkling wines (red, white, rosé)	4.5–7 °C (40–45 °F)
Sweet white wines	7–10 °C (45–50 °F)
White and rosé wines	10–12.5 °C (50–55 °F)
Young light red wines	12.5–15.5 °C (55–60 °F)
Full-bodied red wines	15.5–18 °C (60–65 °F)

— *Desperate situations, desperate remedies* —

When guests call unexpectedly you can chill wines quickly by placing a bottle in a wine bucket containing salt, ice and water. The salt quickly melts the ice and drastically reduces the temperature. Another way is to place the wine in the freezer for about 10 minutes. A cool red wine can be brought to room temperature by being poured into a warmed decanter or by microwaving it for about 45 seconds.

Remember these are *extreme* situations – nothing beats traditional methods of chilling and chambréing wine to a perfect temperature. With wine, perfection is not everything, it is the *only* thing – it takes time.

— *How to put a stop to wasted wine* —

Wine that is left over in a bottle, even when securely corked, becomes lifeless and unpalatable after a couple of days, so down the drain it goes. To prevent this waste a neat piece of equipment exists called a *Vacu-vin* which will reseal an opened bottle of wine, keeping the contents in perfect condition for several days. It extracts the air from the bottle and then reseals it with a special reuseable stopper that preserves the natural freshness of the wine for a longer period, allowing you to drink as much as you like, when you like. There is also a sparkling wine bottle resealer which is particularly useful for Champagne and other bottles of fizz.

STORAGE

Ideally, wine should be stored in a subterranean cellar which has a northerly aspect and is free from vibrations, excessive dampness, draughts and unwanted odours. The cellar should be absolutely clean, well ventilated, with only subdued lighting and a constant cool temperature of 12.5 °C (55 °F) to help the wine develop gradually. Higher temperatures bring wines to maturity more quickly, but this is not preferable.

Table wines should be stored on their sides in bins so that the wine remains in contact with the cork. This keeps the cork expanded and prevents air from entering the wine – a disaster which quickly turns wine to vinegar. White, sparkling and rosé wines are kept in the coolest part of the cellar and in bins nearest the ground (because warm air rises). Red wines are best stored in the upper bins. Commercial establishments usually have special refrigerators or cooling cabinets for keeping their sparkling, white and rosé wines at serving temperature. These may be stationed in the dispense bar – a bar located between the cellar and the restaurant – to facilitate prompt service.

Spirits, liqueurs, beers, squashes, juices and mineral waters are stored upright in their containers, as are fortified wines. The exceptions are port style wines which are destined for laying down.

Only a few people are able to enjoy the benefits of a cellar, so a cupboard or bins under the stairs have to suffice. It is important to locate your wine away from excessive heat – hot water pipes, a heating plant or hot water unit. Heat does far more damage to wine than the cold. Attractive humidity and temperature controlled cabinets are available, but they are expensive.

WINE TASTING

A more apt name for wine tasting is *sensory evaluation* because *sight*, *smell* and *touch* are also involved in the process of appraisal, acceptance or rejection. In the general description below, the following assumptions are made:

(a) that the wine is being tasted before being served to your guests either at home or in a restaurant
(b) that the wine is properly temperatured
(c) that glasses are not only clean but brilliant in appearance.

About 50 ml (2 fl oz) of the wine is poured into the glass, which allows sufficient room for it to be swirled so the bouquet can be enhanced and appreciated.

— *Colour* —

Holding the glass by the stem, tilt it towards the light or a white background and appraise the colour, which should be clear and gleaming, never cloudy, faded or dull.

White wine

These wines range in colour from very pale with green tinges to deep gold. Watch out for any browning at the rim (miniscus) of the wine, as this will reveal that the wine is past its best or has oxidized.

Rosé wine

Such wines can be pink, deep pink or onion skinned in colour, but essentially they should be bright and brilliant in the glass and never subdued-looking or brown.

Red wine

These wines progress from purple in extreme youth, to be ruby, garnet, red and slightly brown or brick red in old age. Reds are usually very intense in colour yet sumptuous in appearance.

— *Smell* —

Smell also known as *nose*, *bouquet* or *aroma*, is very much inter-related with taste. Indeed it is estimated that up to 80 per cent of taste is actually smell. Swirl the wine in the glass. Wine glasses should have lips narrower than the bowl so that as the bouquet evaporates from the wine it becomes more concentrated as it reaches the rim of the glass. Take a good deep smell and, using your imagination, associate the smell with other previously experienced smells (see table below).

Grape	Association
Riesling	apricots, peaches
Chenin Blanc	flowers
Gewürztraminer	tropical fruits, eg lychees
Chardonnay	ripe melon, fresh pineapple
Sauvignon Blanc	gooseberries
Merlot	blackberries
Cabernet Sauvignon	blackcurrants
Pinot Noir	strawberries, cherries, plums (depending on where grown)
Zinfandel	raspberries

— *Taste* —

The taste-character of wine is detected in different parts of the mouth but especially on the tongue – *sweetness* at the tip of the tongue, *acidity* on the upper edges, *saltiness* at the sides and *bitterness* at the back. Take a good sip, roll it round your mouth and, at the same time, suck in some air as this will help to heighten the flavours.

If you are at a wine tasting, make instant notes, otherwise you will forget or later get muddled. *Dryness* or *sweetness* will be immediately obvious, as will *acidity* which, in correct quantities, provides liveliness and crispness. *Astringency* or tannin content associated with some red wines will give a dry coating effect especially on your teeth and gums. *Body*, which is the feel of the wine in your mouth, and *flavour*, the essence of the wine as a drink, will be the final arbiters as to whether you like it or not. *Aftertaste* is the finish the wine leaves on your palate. *Overall balance*, the evaluation of all the above elements, can now be related to price.

When tasting more than five wines it is usual, for obvious reasons, to spit the wine out rather than swallow it. Fortified wines are often appraised by sight and smell alone, as are some spirits which on occasion can even be judged by rubbing a little of the spirit on the palm of the hand.

— *Observations* —

To help you analyze and make judgements on what you are tasting, here are some descriptive words that may be helpful:

Describing colour

Clear, bright, brilliant, gleaming, sumptuous
   dull, grey, hazy, cloudy, pale, faded

White wine Water clear, pale yellow, yellow with green tinges, straw, gold, yellow brown, maderized

Rosé wine Pale pink, orange pink, onion skin, blue-pink, copper

Red wine Purple, brick red, garnet, ruby, tawny, mahogany

Describing nose

Fruity, perfumed, full, deep, spicy, fine, rich, pleasant, nondescript, flat, corky

Describing taste

Bone dry, dry, medium dry, medium sweet, sweet, luscious, unctuous, thin, light, medium, full-bodied, acid, bitter, spicy, hard, soft, silky, smooth, tart, piquant

Summing up

Well balanced, fine, delicate, rich, robust, vigorous, fat, flabby, thick, velvety, harsh, weak, unbalanced, insipid, for laying down, just right, over the hill

It is important that you make up your own mind about the wine. Do not be too easily influenced by the observations of others.

3 Wine producing countries and regions

ALGERIA

Most of the wine produced is either red or rosé and made in three centres: **Alger, Oran** and **Constantine**. The reds are products of the *Carignan, Cinsault* and *Alicante-Bouschet* grapes. The rosés are made from the *Cinsault* and *Grenache*, and the little white that is made comes from *Ugni Blanc* and *Clairette de Provence* grapes.

Traditionally, the wines were consumed at home or used to bulk up blended French wines for sale to the colonies and elsewhere. Since the French pull-out in 1962 quality and trade has diminished, although Côteaux de Mascara, a big, rustic red wine, is still much favoured.

ARGENTINA

Owing to the hot climate and little rainfall, Argentina has had to develop a unique system of irrigation for the vine and other crops to survive. They produce all manner of wines in huge quantities and, until the Falklands trouble, they were making in-roads into the British market especially with the reds.

Mendoza is the biggest and best region. Its output is almost three-quarters of the national total. Other good locations are **San Juan, Río Negro, La Rioja** and **Jujyuy**, part of Salta and Catamarca.

Cabernet Sauvignon, Malbec, Merlot, Pinot Noir, Tempranillo and *Syrah* are the red wine grapes and the *Riesling, Sauvignon Blanc* and *Pinot Blanc* are used to make white wine.

Trapiche, the largest winery in Argentina, makes an exceptionally good Cabernet Sauvignon, as do San Telmo, José Orfila and Bianchi. Crillon makes a still white called Embajador and a sparkling white called Monitor. Proviar makes a good sparkling wine that is marketed as Champania 'M. Chandon'.

AUSTRALIA

It was thought that Australia tipped, as it is, on the edge of Asia, could never successfully compete in the British wine markets against our near and more traditional supply sources. Today, however, almost every wine shop and wine outlet stocks Australian wine and the range gets bigger, annually. The reason, of course, is that Australia produces excellent, good value wines in a great variety of styles, sometimes blending two or more grape varieties to great advantage.

The *Chardonnay*, *Gewürztraminer*, *Rhine Riesling*, *Sauvignon Blanc* and *Sémillon* are the main white grapes, with the brown *Muscat* (Frontignac) for dessert wine. Red wines are made from *Cabernet Sauvignon*, *Pinot Noir*, *Hermitage*, *Shiraz* and *Malbec* grapes.

Labels are informative, stating the grape or combination of grapes used. Sometimes they can be tricky, for example 'C S Malbec' on a label would indicate a blend, in descending proportions, of three red grapes – *Cabernet Sauvignon*, *Shiraz* and *Malbec*.

Generally, however, the labels are easy to understand, as long as they do not give too much technical information such as the *baumé* number (sugar level of the grape when picked), age in cask and date of bottling. Words like 'Private Bin', 'Reserve Bin' and 'Bin Number' may indicate that the wine comes from a single vineyard and is of superior quality.

— *The wines* —

Australian wines are made in **New South Wales** not far from Sydney; in **Victoria** near Melbourne; in the Barossa Valley, Clare Valley, Coonawarra, along the banks of the Murray River and around Adelaide, **South Australia**; in **Queensland** near Brisbane; in **Western Australia**, in the Swan Valley near Perth; and in **Tasmania** near Hobart and a few more places. Below are listed some of the best producers of wine.

Sparkling wines

In Australia wines made by the *méthode traditionnelle* (*méthode Champenoise*) are called Champagne but may not be exported under this name. For special quality try Seppelt's Great Western (VIC), Kaiser Stuhl Special Reserve (SA) and Lindeman's Grand Imperator (NSW).

Others are:

Angas Brut (SA)
McWilliams Brut (NSW)
Mildara Yellowglen (VIC)

Yalumba Brut de Brut (SA)
Tulloch Brut (NSW)

White wine

Enterprise Rhine Riesling (SA)
Henschke Rhine Riesling Spätlese (SA)
Leo Buring Rhine Riesling (SA)
Petaluma Rhine Riesling (SA)
Rosemount Estate Rhine Riesling (NSW)
Lindeman's Hunter River Riesling (NSW)
Château Tahbilk Riesling (VIC)
Quelltaler Grande Reserve Riesling (SA)
Leeuwin Estate Chardonnay (WA)
Rosemount Estate Chardonnay (SA)
Petaluma Chardonnay (NSW)
Drayton's Chardonnay (NSW)
Lake's Folly Chardonnay (NSW)

Rothbury Estate Chardonnay (NSW)
Saxonvale Chardonnay (NSW)
Tyrrell's Chardonnay (NSW)
Lindeman's Hunter River White Burgundy (NSW)
Wynn's Huntersfield White Burgundy (SA)
Lindeman's Hunter River Porphyry Sauternes (NSW)
McWilliams Mount Pleasant Sauternes (NSW)
Rosemount Estate Sauvignon Blanc (NSW)
Henschke Sémillon (SA)
Hill-Smith Estate Sémillon (SA)
Kaiser Stuhl Sémillon (SA)

Reds

Lindeman's Hunter River Burgundy (NSW)
Lindeman's Cawarra Claret (NSW)
Penfolds St Henri Claret (SA)
Yalumba Signature Blend Claret (SA)
Reynella Coonawarra Cabernet Shiraz (SA)
Saltram Mamre Brook Cabernet Shiraz (SA)
Seppelt Cabernet Shiraz (SA)
Petaluma Cabernet Shiraz (SA)

Penfolds Cabernet Shiraz (SA)
Wolf Blass Black Label Cabernet Shiraz (SA)
Brokenwood Hermitage Cabernet Sauvignon (NSW)
Penfolds Grange Hermitage (SA)
Tulloch Glen Elgin Estate Hermitage (NSW)
Capel Vale Shiraz (WA)
Pipers Brook Pinot Noir (TAS)
Moorilla Estate Pinot Noir (TAS)

The following produce high quality Cabernet Sauvignon wines:

Château Reynella (SA)
Château Tahbilk (VIC)
Laira (SA)
Hardy's Reserve Bin (SA)
Penfold (SA)
Redmann (SA)
Stanley Leasingham (SA)
Taltarni (VIC)

Rothbury Estate (NSW)
Wynns Coonawarra Estate (SA)
Brown Bros Milawa (VIC)
Lake's Folly (NSW)
Seppelt Reserve Bin (SA)
Evans & Tate (WA)
Robinsons Family (Qld)

Australian style sherry

Mildara George (VIC)
Quelltaler Granfiesta Dry Pale (SA)

Yalumba Chiquita (SA)

Australian style port

Hardy's Vintage (SA)
Kaiser Stuhl Jubilee Port (SA)
Lindeman's Macquarie (NSW)
Reynella Vintage (SA)

Seppelt Para (SA)
Yalumba Galway Pipe (SA)
Elsinore Vintage (Qld)

AUSTRIA

About three-quarters of Austrian wine is white, made mostly from the native grape *Grüner Veltliner*, with contributions also from the *Rhine Riesling*, *Welsch Riesling*, *Weissburgunder (Pinot Blanc)*, *Gewürztraminer*, *Müller-Thurgau* and *Muskat-Ottonel*. Red wines when made, come from the *Blauer Spätburgunder (Pinot Noir)*, *Blaufränkisch*, *Portugieser* and *Saint Laurent* grapes.

Since the infamous diethylene glycol wine scandal of 1985 a few perpetrators have caused substantial damage to the Austrian wine trade, which now shows a shortfall of 80 per cent of the export market. Let us hope that the reputable producers can withstand the inevitable financial stresses and strains, and that their good, stylish wine will soon reach the traditional markets as before.

The most popular white wine in Austria is Gumpoldskirchner made in a village near **Baden**, south of Vienna from special grapes not mentioned above. They are *Rotgipfler* and *Spätrot*, blended in equal proportions. Further south of Vienna the really good red, Vöslauer, is made in the village of **Bad Vöslau**. Vineyards to the west of Vienna especially in the **Wachau** district produce fine white such as Dürnsteiner Katzensprung, Dürnsteiner Flohaxen, Lóiben Kaiserwein, Riede Lóibenberg, Kremser Kögl and Kremser Wachtberg.

Burgenland in the eastern part of Austria gets lots of sunshine and has ideal conditions for producing overripe grapes, which in extreme ripeness are known as *ausbruch*. A delicious example of this is the wine from Neusiedler See called Rust. Other good wines are the white styles Neuberger, Mörbischer, St Georgener Welschriesling and the light and fruity red Blaufränkischer.

In **Steiermark** in the southern corner of Austria they make a lovely blush style wine called Schilcher, which results from very brief maceration of the grape skins. In West Steiermark the speciality is the onion-skin-coloured rosé called Zwiebelschilcher.

— *General wine styles* —

Bergwein	mountain vineyard wine
Hcurigerwein	new wine, from the Vienna Woods district; released for sale on 11 November each year and sold by the jugful in bars (*Heurigen*)
Perlwein	semi-sparkling or *spritzig* wine
Reidwein	single vineyard wine
Schaumwein	fully sparkling wine such as Schlumberger Blanc de Blancs, made by the *méthode Champenoise*
Schluckwein	(gulping wine), a thirst quencher
Schoppenwein	(swilling wine), a thirst quencher

Viticulturists in many parts of the world are indebted to Dr Lenz Moser, of Austria, who devised the high and wide method of vine cultivation, which is especially associated with white wine production. He introduced the practice of training the vines high and well spaced out so that tractors could be used to farm the vines.

BULGARIA

Moslem austerity frowned on the cultivation of the vine in Bulgaria for many centuries. In 1944, the government recognized that wine could be a valuable export and encouraged the planting of vineyards, introducing best quality grapes and modern methods of viticulture and vinification.

Although Bulgaria is only fourteenth in the table of producers it is actually the fourth largest exporter. The wines are very good and reasonably priced. They are matured in oak cask – whites for up to 18 months and the reds for three years. Whites to look out for are Dimiat, Riesling, Chardonnay, Sauvignon Blanc and Misket, and the reds Gamza, Mavrud, Melnic and Cabernet Sauvignon are all generous, soft and rounded in flavour.

CANADA

Ontario is by far the biggest and most important wine region producing some 85 per cent of Canada's total. Much of the wine comes from the *Vitis labrusca* family (*Concorde*, *Catawba*, *Delaware*, *Niagara* and *President*). In more recent times the *Vitis vinifera* varietals such as *Aligoté*, *Chardonnay*, *Gewürztraminer* and the *Johannisberg Riesling* for whites and the *Pinot Noir* and *Gamay* for reds have been cultivated.

The major producers are Calona Wines Ltd, Andres Wines, Brights Wines, Château des Charmes, Château-Gai Wines, Colio Wines, Inniskillin Wines (one of the best names), London Winery, Montravin Cellars, Reif Winery Inc and Willowbank Estate.

Besides the red and white table wines, dessert style wines including Eiswein, and a substantial amount of sparkling wine are made.

CHILE

Although Chile was one of the few places to dodge the *Phylloxera vastatrix* scourge, they restocked their vineyards in the mid nineteenth century with the finest of the *noble* varietal European vines. The resulting improvement in quality has elevated the red wines, especially the Cabernet Sauvignons, to world class level. Look out for this style from Concha y Toro, Cousino-Macul, Miguel Torres and Santa Rita. They are full, rich and velvet smooth, certain to please both palate and pocket. The best Chilean wines come from the **Maipo** and **Aconcagua** valleys.

Reds are also produced either as blenders, or in their own right, from *Cabernet Franc*, *Malbec*, *Merlot* and *Pinot Noir* grapes. Torres also produces a good rosé from the *Cabernet Sauvignon* grape and a refreshing sparkling white made by the *méthode traditionnelle*. The still whites have not, as yet, reached the same high standards as the reds but again Torres Sauvignon Blanc has good balance and elegance. The *Chardonnay* and *Sémillon* grapes are also used to produce sound if not spectacular whites.

— *Label language* —

Envasado en Origen:	estate bottled
Viñas Courant:	one year old
Viñas Special:	two years old
Viñas Reserva:	four years old
Viñas Reservado:	six years old

CHINA

China makes some rice 'wine' and white grape wines called Dynasty, Heavenly Palace and Great Wall. A sweet wine called Meikuishanputaochu is also made, as is a reasonable red wine called Cabernet d'Est. Climatic conditions are against quality.

CYPRUS

The sunshine island of Cyprus produces a glistening selection of wines running from dry, medium and the more popular sweet sherries, to the renowned dessert wine, Commandaria. Although the sherries are the more popular, Commandaria, reckoned to be the oldest known wine in the world, has more distinction. Made by farmers in 11 villages on the southern slopes of the **Troodos** mountains, the grapes are left to dry on roadsides and on rooftops for 10–15 days. They are then pressed and the new wine is kept until spring when it is sold to the wine establishments at **Paphos** and **Limassol**. Here it is flavoured with cloves, resin, scented wood, and fortified with local brandy.

Commandaria has a reputation for promoting longevity and is drunk locally as a tonic – some excuse! All we know is that this honey-sweet, amber-red wine is delicious after a meal. Try Commandarie St John. The island also produces good table wines.

Light whites

Amathus, Arsinoe, Palomino

Medium whites

Aphrodite, Bellapais (white and rosé semi-sparkling) and a newcomer, Thisbe, which has a nice nose and a smooth flavour

Sweet whites

Hirondelle, St Hilarion and St Panteleimon

Rosé wine

The best is Coeur de Lion; the newish Amorosa is also good, as is Kokkineli, a deep rosé wine

Red wine

Of the reds Kykko, Olympus and Salamis are nicely mellow, Agravani is dry and flavoursome, whilst Afames, Kolossi and Othello are full bodied and fragrant

Sparkling wine

Of the sparkling wines two that are agreeable and easy to remember are Avra and Duc de Nicosie. Watch for the names of the following wineries on Cyprus labels: Keo, Etko, Sodap and Loel

ENGLAND

The English have a hang up about their wines. For some reason, average wine merchants seem reluctant to promote or even feature

English wines on their lists, whilst the bigger and more established ones will offer only one example, perhaps two at the most. Likewise with restaurant wine lists, most are devoid of English wines, and those that do show some often discourage sales by unreasonable mark ups. This is all a great pity because English wine is good, if not great, and accompanies lighter food dishes more than adequately. 'English' wines are made from grapes grown in England alone. Wines labelled 'British' are made from imported unfermented grape juice (*must*) and may not be called 'English' wines.

The vine was first introduced into Britain by the Romans, and the 1086 Domesday Book survey recorded the existence of 83 vineyards. This number slowly increased, most being in the hands of monasteries, religious orders or great houses. When Eleanor of Aquitaine married Henry II in 1152 part of her dowry was the lands of Bordeaux which England owned thereafter for some 300 years. The red wine of Bordeaux was markedly superior to the wines of England and this fact, together with the dissolution of the monasteries in the 1530s, resulted in many vineyards falling into disuse.

It was not until well after the end of the Second World War that the English wine industry resurfaced physically and commercially. Today in England and Wales there are more than 300 vineyards occupying over 1,000 acres, producing mainly white wines from Germanic vine strains. These strains are especially suited to cold, northerly vineyards, with the extreme limit of cultivation being a rough line across the country from The Wash. The cold climate is a great problem, as are the wet summers such as in 1984 when a third of the crop was lost. The English winegrowers, as needs be, are a hardy and stubborn lot. Their fortitude is occasionally rewarded with a great vintage like that of 1989 when 3.2 million litres were produced (4.26 million bottles).

Experience has shown that the following vines produce the best white wine: *Müller-Thurgau, Schönburger, Ortega, Reichensteiner, Huxelrebe, Ortega, Bacchus, Gütenborner, Seyve Villard, Morio Muscat, Riesling* and *Sylvaner*, and a French hybrid, *Seyval Blanc*.

There are also red and rosé wines made from *Pinot Noir, Zweigeltrebe* and *Gamay* grapes, but so far they have not developed beyond the ordinary.

Look out for the *EVA* quality seal which is awarded annually by the English Viticultural Association to wines that are submitted for official testing. The EVA was formed in 1965 and their seal on a bottle indicates quality.

— *Major English vineyards* —

Adgestone (Isle of Wight)
Ascot (Berkshire)
Barton Manor (Isle of Wight)
Beaulieu (Hampshire)
Biddenden (Kent)
Bothy (Oxfordshire)
Breaky Bottom (Sussex)
Bruisyard (Suffolk)
Carr Taylor (Sussex)
Cavendish Manor (Suffolk)
Chilford Hundred (Cambridgeshire)
Chilsdown (Sussex)
Ditchling (Sussex)
Elmham Park (Norfolk)

English Wine Centre (Sussex) – sells a
    range of English wines
Hambledon (Hampshire)
Lamberhurst (Kent)
Merrydown (Sussex)
Pilton Manor (Somerset)
Tenterden and Spots Farm (Kent)
Staple (Kent)
Three Choirs (Gloucestershire)
Wellow (Hampshire)
Westbury (Berkshire)
Wootton (Somerset)
Wraxall (Somerset)

FRANCE

The glorious wine gardens of France produce a diversity of wine styles, generally of noble quality. Besides the excellent natural aspects of soil and climate, quality is controlled at all stages of production.

The making and labelling of French wine is now governed by EEC wine laws and defined as follows:

— *Vins de Qualité produits dans des regions determinées (VQPRD)* —

These are quality wines produced from grapes grown in specific regions. They are subdivided into two categories:

Vins d'appellation d'origine contrôlée (AOC)

This labelling guarantees:

(a) Area of production
(b) Grape varieties used
(c) Pruning and cultivation
    methods
(d) Maximum yield per hectare
(e) Minimum alcohol content
(f) Methods of vinification and
    preservation.

FRANCE

N

Reims

Paris   Épernay   Strasbourg

CHAMPAGNE   *ALSACE*

VAL-DE-LOIRE   Auxerre   *CHABLIS*

Nantes •Anjou   *VOUVRAY*   Sancerre   Dijon
        SAUMUR   Tours   Pouilly-sur-
                TOURAINE   Loire

                        BOURGOGNE *JURA*

COGNAC   Lyon

MÉDOC   *SAVOIE*
BORDEAUX
        MONBAZILLAC   *CÔTES*
GRAVES   *DU RHONE*
    SAUTERNES
                            Avignon

ARMAGNAC   *CÔTES DE*
                                PROVENCE
            MIDI
        CORBIÈRES   Marseille
        LANGUEDOC
            ROUSSILLON

Vins délimités de qualité supérieure (VDQS)

These are wines of superior quality produced in delimited areas with
the following conditions guaranteed:

(a) Area of production
(b) Grape varieties used
(c) Minimum alcohol content
(d) Methods of viticulture and vinification.

Although the wines have to be good to merit the VDQS label, they are less fine than the AOC wines.

— *Vins de table* —

The second labelling category, Vins de table, is also divided into two:

Vins de pays (VP)

Local or country wine. Medium in quality, these wines must be made from recommended grapes grown in a certain area or village. They must have a minimum alcohol content, and come from the locality stated on the label.

Vins de consommation courante (VCC)

Wines for everyday consumption and sold by the glass, carafe or *pichet* in cafés and bars all over France. Often completely authentically French, these wines may also be blended with other EEC wines of similar style. Non-EEC wines may not be blended with French wines.

— *Label language* —

Appellation d'origine contrôlée (AOC)	associated with the best French wines and guarantees origin of wine named on the label
Crémant	sparkling wine, but not Champagne
Cru	(growth) – used to describe a single vineyard. *Grand cru* and *premier grand cru* indicate higher and highest quality vineyards
Cuvée	blend or contents of a vat of wine
Château/domaine	estate
Cuve close	sparkling wine made in bulk inside a sealed vat
Millesime	vintage date
Mis en bouteille au château/domaine	bottled at the estate

Mis en bouteille dans nos caves/chais	bottled in our cellars, usually by a large wine company. Not estate bottled
Mis en bouteille à la propriété par	bottled at the property or estate for somebody else, usually a wine dealer
Méthode Champenoise	Champagne method used in making some sparkling wines. Also known as *méthode traditionnelle*
Moelleux	sweetish and smooth
Mousseux	sparkling
Négociant	wine handler who buys bulk wine from growers and sells it under his own label
Pétillant	lightly sparkling
Propriétaire, Récoltant	owner, grower
Récolte . . .	harvested . . . – followed by the vintage year
Sec	dry (*demi-sec*: medium dry and *doux*: sweet)
Sur lie	wine left to mature on its lees (sediment) before being bottled
Vendange tardive	late harvested grapes which produce sweeter wine

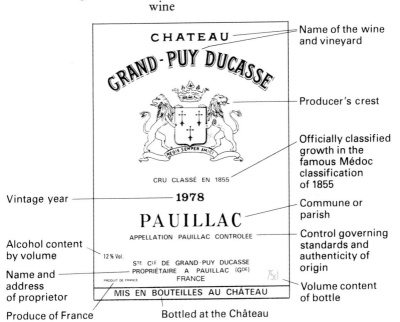

CHATEAU — Name of the wine and vineyard

GRAND-PUY DUCASSE

— Producer's crest

CRU CLASSÉ EN 1855 — Officially classified growth in the famous Médoc classification of 1855

1978 — Vintage year

PAUILLAC — Commune or parish

APPELLATION PAUILLAC CONTROLEE — Control governing standards and authenticity of origin

12 % Vol. — Alcohol content by volume

S^{TE} C^{IE} DE GRAND-PUY DUCASSE
PROPRIÉTAIRE A PAUILLAC (G^{DE})
FRANCE — Name and address of proprietor

PRODUIT DE FRANCE — Produce of France

75cl — Volume content of bottle

MIS EN BOUTEILLES AU CHÂTEAU — Bottled at the Château

143

Vin de pays	a local wine usually of sound quality
Vin de table	*Vin ordinaire* – ordinary table wine
Vin délimité de qualité supérieure	(VDQS) wine of superior quality produced in the region as stated on the label
Vin doux naturel	(VDN) – sweet wine whose fermentation has been muted by the addition of alcohol
Vin Primeur	(*Vin de l'année, Vin nouveau*) wine made to be drunk within a year

— *Alsace* —

Alsace wines are mainly white after the style of German Rhine wines and are marketed in tall green *flûte d'Alsace* bottles under grape rather than place names. Alsace Riesling is probably the most popular wine but other grapes used are *Gewürztraminer, Pinot Gris* (Tokay), *Pinot Blanc* (Klevner), *Muscat* and *Sylvaner*. The *Pinot Noir* is used to make a little red and a rosé known as Clairet d'Alsace or Schillerwein. There is also a white wine, Alsace Edelzwicker, made from a blend of noble grapes such as *Riesling, Sylvaner* etc. The very pleasant sparkling Crémant d'Alsace is made mostly from a combination of *Riesling* and *Pinot* grapes.

Alsace wines have variety in style and flavour and are reliably well made. The region is also famous for its Alcool Blanc, notably Eau-de-Vie de Poire William.

— *Bordeaux* —

Ever since 1152, when Henry II of England married Eleanor of Aquitaine and got Gascony and most of south west France as a dowry, the wines of Bordeaux have usually held pride of place in an Englishman's cellar. Until 1453 the area was a part of the English crown and Claret, the red wines of Bordeaux, quickly became popular with Royal patronage.

Some 70 per cent of the total production are AOC quality of which two-thirds are red and one-third white. The red wines are made from the *Cabernet Sauvignon, Cabernet Franc, Merlot, Malbec* and *Petit-Verdot* grapes and the whites from the *Sémillon, Sauvignon Blanc* and *Muscadelle* grapes.

BORDEAUX

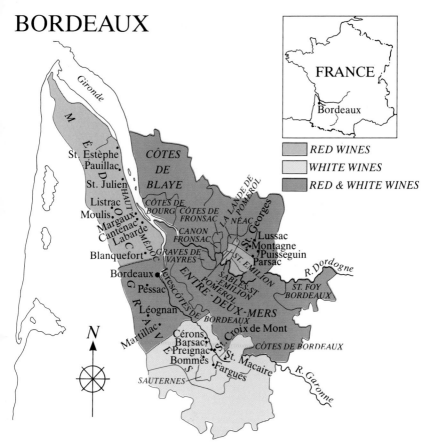

The main wine growing areas are: **Médoc, Saint-Emilion, Pomerol, Entre-Deux-Mers, Graves, Cérons, Sauternes**, and **Bourg, Blaye** and **Fronsac**.

Médoc

The most famous area in the world for the production of quality, long lasting red wines. The area is so renowned for fine wines that in 1855 61 of the better wines were classified into five divisions known as *crus*, or growths, and that classification, with just a few exceptions, holds good to this day.

GRANDS CRUS CLASSÉS OF THE MEDOC
The official classification of 1855

	Vineyard	Commune
First Growths (*Premiers Crus*)	Ch. Lafite-Rothschild	Pauillac
	Ch. Latour	Pauillac
	Ch. Margaux	Margaux
	Ch. Mouton-Rothschild	Pauillac[1]
	Ch. Haut-Brion	Pessac[2]
Second Growths (*Deuxièmes Crus*)	Ch. Rausan-Ségla	Margaux
	Ch. Rauzan-Gassies	Margaux
	Ch. Léoville-Lascases	St Julien
	Ch. Léoville-Poyférré	St Julien
	Ch. Léoville-Barton	St Julien
	Ch. Dufort-Vivens	Margaux
	Ch. Gruaud-Larose	St Julien
	Ch. Lascombes	Margaux
	Ch. Brane-Cantenac	Cantenac
	Ch. Pichon-Longueville-Baron	Pauillac
	Ch. Pichon-Longueville-Lalande	Pauillac
	Ch. Ducru-Beaucaillou	St Julien
	Ch. Cos-d'Estournel	St Estèphe
	Ch. Montrose	St Estèphe
Third Growths (*Troisièmes Crus*)	Ch. Kirwan	Cantenac
	Ch. d'Issan	Cantenac
	Ch. Lagrange	St Julien
	Ch. Langoa Barton	St Julien
	Ch. Giscours	Labarde
	Ch. Malescot-Saint-Exupéry	Margaux
	Ch. Boyd-Cantenac	Cantenac
	Ch. Palmer	Cantenac
	Ch. La Lagune	Ludon
	Ch. Desmirail	Margaux

[1] Upgraded from the second to the first growth in 1973
[2] Graves District

	Ch. Cantenac-Brown	Cantenac
	Ch. Calon-Ségur	St Estèphe
	Ch. Ferrière	Margaux
	Ch. Marquis-d'Alesme-Becker	Margaux
Fourth Growths (*Quatrièmes Crus*)	Ch. Saint-Pierre-Sevaistre	St Julien
	Ch. Talbot	St Julien
	Ch. Branaire-Ducru	St Julien
	Ch. Duhart-Milon-Rothschild	Pauillac
	Ch. Pouget	Cantenac
	Ch. La Tour-Carnet	St Laurent
	Ch. Lafon-Rochet	St Estèphe
	Ch. Beychevelle	St Julien
	Ch. Prieuré-Lichine	Cantenac
	Ch. Marquis-de-Terme	Margaux
Fifth Growths (*Cinquièmes Crus*)	Ch. Pontet-Canet	Pauillac
	Ch. Batailley	Pauillac
	Ch. Haut-Batailley	Pauillac
	Ch. Grand-Puy-Lacoste	Pauillac
	Ch. Grand-Puy-Ducasse	Pauillac
	Ch. Lynch-Bages	Pauillac
	Ch. Lynch-Moussas	Pauillac
	Ch. Dauzac	Labarde
	Ch. Mouton Baronne Philippe	Pauillac
	Ch. du Tertre	Arsac
	Ch. Haut-Bages-Libéral	Pauillac
	Ch. Pédesclaux	Pauillac
	Ch. Belgrave	St Laurent
	Ch. Camensac	St Laurent
	Ch. Cos-Labory	St Estèphe
	Ch. Clerc-Milon-Rothschild	Pauillac
	Ch. Croizet-Bages	Pauillac
	Ch. Cantemerle	Macau

Saint-Emilion

Another famous red wine district was classified in 1955 and revised in 1969 and 1985. The wines are full and rounded with Château Ausone and Château Cheval Blanc being fine examples.

THE 1955 CLASSIFICATION
First Great Growths (*Premiers Grands Crus*)

Ch. Ausone Ch. Cheval-Blanc } Grade A	Ch. Beauséjour-Duffau-Lagarrosse Ch. Beauséjour-Becot Ch. Bel-Air Ch. Canon Ch. Figeac Clos Fourtet Ch. La Gaffelière Ch. La Magdelaine Ch. Pavie Ch. Trottevieille	} Grade B

Pomerol

A neighbouring district, to the west of Saint Emilion, produces rich meaty but smooth reds; Château Pétrus is the best known. These wines were never officially classified but some of the better ones are listed below:

Cru Exceptionnel	Château Pétrus
Other Principal Crus	Ch. Certan de May Ch. Certan-Giraud Ch. la Conseillante Ch. l'Eglise-Clinet Ch. l'Evangile Ch. la Fleur Pétrus Ch. la Grave-Trigant-de-Boisset Ch. Latour à Pomerol Ch. Petit-Village Ch. le Pin Ch. Trotanoy Clos de l'Eglise (Moreau) Vieux Ch. Certan

Entre-Deux-Mers

A stretch of land between the two rivers Garonne and Dordogne produces decent reds plus dry and sweet white wines. AOC is at present limited to the dry white wines; the others may be sold as Bordeaux or Bordeaux Supérieure.

Graves

This district gets its name from the gravel content of the soil and was classified in 1959. Both red and white wines are made, although it is the reds that have the better reputation, especially the outstanding Château Haut-Brion which was nominated as a *premier cru* (first growth) wine in the famous 1855 classification of the Médoc.

CLASSIFIED GROWTHS OF THE GRAVES		
Red wines *(classified in 1953 and confirmed in 1959)*	Ch. Bouscaut	Cadaujac
	Ch. Haut-Bailly	Léognan
	Ch. Carbonnieux	Léognan
	Domaine de Chevalier	Léognan
	Ch. Fieuzal	Léognan
	Ch. Olivier	Léognan
	Ch. Malartic-Lagravière	Léognan
	Ch. La Tour-Martillac	Martillac
	Ch. Smith-Haut-Lafitte	Martillac
	Ch. Haut-Brion	Pessac
	Ch. La Mission-Haut-Brion	Pessac
	Ch. Pape Clément	Pessac
	Ch. Latour-Haut-Brion	Talence
White wines *(classified in 1959)*	Ch. Bouscaut	Cadaujac
	Ch. Carbonnieux	Léognan
	Domaine de Chevalier	Léognan
	Ch. Olivier	Léognan
	Ch. Malartic-Lagravière	Léognan

Cérons

This area is on the borders of Graves, producing white fragrant wines which vary in style both from dry to extremely sweet.

Sauternes

This district produces the most remarkable naturally sweet, golden wine you can taste. It is due to a phenomenon called *Botrytis cinerea*, also known as *pourriture noble* or noble rot.

In autumn the local climatic conditions of morning moisture followed by strong heat create the humidity that encourages spores to form and fester on the outside of the grape skins. The resulting fungus feeds on the moisture within each grape, gradually reducing the contents by two-thirds and, at the same time, concentrating the remaining juices into a rich sugar syrup. The dehydrated grapes are

picked only when they have reached a specific state of rotting, so the vineyards are gone over time and time again. The resulting wine, luscious with creaminess and intense sweetness, has a vigour about it that stands it apart from other sweet wines. This quality is best exemplified by the famous Château d'Yquem.

Also within the limitations of the **Sauternes** boundaries are the regions of **Barsac**, **Bommes**, **Fargues** and **Preignac**, all producing very good sweet wines.

SAUTERNES AND BARSAC
The 1855 classification

First Great Growth (*Premier Grand Cru*)	Ch. d'Yquem	Sauternes
First Growths (*Premiers Crus*)	Ch. La Tour-Blanche	Bommes
	Ch. Lafaurie-Peyraguey	Bommes
	Clos Haut-Peyraguey	Bommes
	Ch. Rayne-Vigneau	Bommes
	Ch. Suduiraut	Preignac
	Ch. Coutet	Barsac
	Ch. Climens	Barsac
	Ch. Guiraud	Sauternes
	Ch. Rieussec	Fargues
	Ch. Rabaud-Promis	Bommes
	Ch. Sigalas-Rabaud	Bommes
Second Growths (*Deuxièmes Crus*)	Ch. de Myrat	Barsac*
	Ch. Doisy-Daëne	Barsac
	Ch. Doisy-Dubroca	Barsac
	Ch. Doisy-Védrines	Barsac
	Ch. d'Arche	Sauternes
	Ch. Filhot	Sauternes
	Ch. Broustet	Barsac
	Ch. Nairac	Barsac
	Ch. Caillou	Barsac
	Ch. Suau	Barsac
	Ch. de Malle	Preignac
	Ch. Romer	Fargues
	Ch. Lamothe	Sauternes

* No longer in production

Bourg, Blaye and Fronsac

These are three areas producing white and red wines, but they are probably best known for their bright, full-bodied robust reds of *Cru Bourgeois* quality.

— *Burgundy* —

The vineyards of Burgundy stretch from Chablis in the far north to Lyon in the south, producing in good years white and red wines of excellence. However, in some years the grapes in the more northerly limitations do not ripen properly and man takes over from nature to add sugar to the must in order to bring the alcohol content up to that of a similar wine produced in a good year. This doctoring of the wine is legal and is known as *chaptalization* after Dr. Chaptal (1756–1832) who first introduced the practise. Chaptalized wines cannot be sold as vintage wines – they do not even bear that stamp of class and are often sold as second name wines or co-operative wines. This means that a vintage on a Burgundy label really means something.

Before 1789 most of the vineyards belonged to the Church but in the French Revolution they were seized and fragmented into saleable lots or *climats* which local farmers could afford to buy. Multi-ownership of original vineyards has been a tradition ever since; in fact, one famous vineyard, Clos de Vougeot, is owned by 85 growers, each entitled to sell his wine by the vineyard name. With the sugaring of the must and the dissecting of vineyards, the buying of Burgundy can at times be something of a gamble!

Owing to their popularity in countries such as America, Belgium and Great Britain, there is a perpetual shortage of the finest Burgundies (which are sold at very high prices). Many of the finest wines are domaine bottled by the grower and sold under the vineyard or *commune* (parish) label. Much is also sold to *négociants* who prepare a blend of several wines thereby averaging the quality as well as the price. It is therefore advisable to acquaint oneself with the names of reliable *négociants* or shippers such as Patriarche, Bouchard Père et Fils, Joseph Drouhin, Georges Duboeuf, Louis Jadot, and Louis Latour when considering Burgundy.

Of all the wine produced, five-sixths is red and only one-sixth white. The great reds are made from the classic *Pinot Noir* grape and others from the *Gamay* or *Passe-tout-grains* (a mixture of one-third *Pinot*

Noir and two-thirds *Gamay*). The excellent whites are produced from *Chardonnay* grapes and the less fine from the *Aligoté*.

Burgundy is divided into six regions: **Chablis, Côte de Nuits, Côte de Beaune, Côte Chalonnaise, Côte Mâconnaise** and **Beaujolais**.

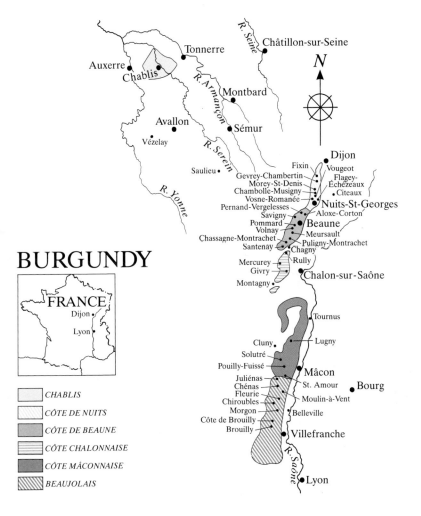

BURGUNDY

CHABLIS

CÔTE DE NUITS

CÔTE DE BEAUNE

CÔTE CHALONNAISE

CÔTE MÂCONNAISE

BEAUJOLAIS

Chablis

The flinty dry white wines coming from this district have attained the greatest possible distinction, being generally regarded as the best accompaniment to shellfish and light delicate foods. Going from basic to brilliant, the wines are classified as *Petit Chablis*, *Chablis*, *Chablis Premier Cru* and *Chablis Grand Cru*. The latter has seven vineyards of the highest calibre:

Vaudésir	Grenouilles	Blanchots
Les Preuses	Les Clos	Les Bougros
Valmur		

Côte de Nuits

A red wine district of great renown producing full bodied meaty wines which develop gradually into silky smooth wines of exceptional class. Some examples of the famous *communes* and outstanding vineyards are shown in the table below.

Commune	*Vineyard*
Gevrey-Chambertin	Chambertin
	Clos de Bèze
Morey Saint-Denis	Clos de Tart
	Clos Saint-Denis
Chambolle Musigny	Musigny
	Les Bonnes-Mares
Vougeot	Clos de Vougeot
Flagey-Echézeaux	Echézeaux
	Grands Echézeaux
Vosne Romanée	La Romanée-Conti
	Le Richebourg
	La Romanée Saint-Vivant
Nuits-Saint-George	Les Saint-Georges
	Clos de la Maréchale

Côte de Beaune

This district is famous for fine, but less assertive, supple reds which age in a reasonably short time. Quality white wines are also produced.

Côte de Beaune reds

Commune	Vineyard
Pernand-Vergelesses	Les Vergelesses
Aloxe Corton	Le Corton
	Corton Clos du Roi
Savigny-les-Beaune	Les Lavières
Beaune	Les Marconnets
	Le Clos des Mouches
Pommard	Les Epinots
	Les Rugiens
Volnay	Les Caillerets
	Les Champans
Santenay	Les Gravières
	Le Clos de Tavannes
Chassagne-Montrachet	Clos Saint-Jean

Côte de Beaune *can be a combination wine from the Beaune area.*
Côte de Beaune Villages *comes from one or more villages which have a right to the appellation.*

Côte de Beaune whites

Commune	Vineyard
Meursault	Les Perrières-Dessous
	Les Charmes-Dessous
Puligny-Montrachet	Le Montrachet
Chassagne-Montrachet	Le Chavalier Montrachet
	Le Batard Montrachet (overlaps both communes)

Côte de Beaune whites are celebrated for their superb style and quality, the finest being produced in the Montrachet and neighbouring Meursault vineyards. However, there are also other fine wines from Côte de Beaune, eg Corton-Charlemagne (Aloxe Corton).

Beaune is also known for its *Hospices de Beaune*, a fifteenth century almshouse which looks after pensioners and the poor. The hospital was founded in 1443 by the then Chancellor Nicolas Rolin who endowed the premises with vineyards and encouraged others to be benefactors in like manner. The sale of wines from these vineyards supports the Hospices and, since 1859, on the third Sunday of each November a famous wine auction takes place in public, this is its chief source of income. Each lot of wine is auctioned *à la chandelle*: the bidding starts when the auctioneer lights a candle, which is snuffed when the lot is sold. Another candle is then lit to start the bidding for the next lot. Usually just two candles are used. The prices realised, although benevolently generous, set a standard for Burgundy prices for that particular year.

> *The* **Côte de Nuits** *and* **Côte de Beaune** *are together popularly known as the* **Côte d'Or** *(Golden Hillsides), not only because of the beautiful vista of their gold coloured vineyards in the autumn but also because of the wealth that the wines have generated.*

Côte Chalonnaise

This district produces lighter red wines which mature quickly but lack the grandeur of the Côte d'Or wine. Mercurey, Rully and Givry are the best examples while Montagny, made exclusively from *Chardonnay* grapes, and Bouzeron are the best of the white wines made.

Côte Mâconnaise

This region is a prolific producer of light, red, fruity wines – pleasant wines for picnics, wine and cheese parties and the like. They are usually sold as Mâcon Rouge or Mâcon Supérieur and the latter must have an alcohol content of 11%.

By far the best Mâconnais wine is the white Pouilly Fuissé made from *Chardonnay* grapes in the *communes* of Fuissé, Salutré, Vergisson and Chaintré. Adjoining *communes*, Pouilly-Vinzelles and Pouilly-Loché, also produce these typically fine fresh and vigorous wines while the appellation Mâcon-Villages (Blanc) has sound whites such as Mâcon Lugny, Mâcon Prissé and Mâcon Clessé.

Saint Véran, made in vineyards which overlap Mâcon and Beaujolais, is similar in style.

Beaujolais

Although some good white wine is made in Beaujolais, it is the light, fruity aromatic reds that give the area fame and fortune. Most wines are sold under the parish name of origin (*Beaujolais Cru*), ie:

Saint-Amour	Juliénas
Chénas	Chiroubles
Morgon	Moulin-à-Vent
Brouilly and Côte de Brouilly	Fleurie

Genuine Beaujolais is made in huge quantities from *Gamay* grapes and ranges in quality from basic *Beaujolais*, *Beaujolais Supérieur* to *Beaujolais Villages* (a blend of wine from two villages or more).

The vogue for *Beaujolais Nouveau* (Beaujolais Primeur) accounts for half the Beaujolais sales. This light but pleasant swilling wine is made by a method called *macération carbonique*. The whole bunches, together with their stalks, are piled in a closed vat or container and the grapes are left to press themselves. As the weight of the grapes release the juice, natural fermentation begins to take place within each grape. Later the process is completed according to local tradition resulting in a wine which is soft, fragrant, fruity and ready for drinking once bottled. It is released for sale on the third Thursday in November and is best drunk between then and Christmas. Beaujolais de l'Année is somewhat different being offered for sale within a year of its vintage.

Beaujolais is at its most refreshing when drunk at cellar temperature, ie slightly chilled.

— *Champagne* —

Champagne is a protected name and is unquestionably the greatest of all sparkling wines. The vineyards are located in north east France and cover some 21,000 hectares (51,870 acres). The three demarcated areas are **Montagne de Reims**, **Vallée de la Marne** and **Côte des Blancs** (the white hillsides near Épernay which are planted with the white *Chardonnay* grape, hence the name).

Besides *Chardonnay*, which gives an attractive crispness, delicacy and finesse to Champagne, two black grapes also used – the *Pinot Noir* and the *Pinot Meunier*. These give body and balance to the wine. When the wine is made solely from the *Chardonnay* grape it is known as *Blanc de Blancs* (white of whites). The resulting wine is very light, refreshing and delicate. If made only from black grapes it is called *Blanc de Noirs* and results in a much fuller, rounded, heavier Champagne. The great majority of Champagne is made from a combination of all three grapes which provides the beautiful unique balance that we all appreciate.

Most vineyards have a predominance of chalky soil which readily absorbs moisture to keep vine roots healthy. They are graded from 100 per cent (*Grands Crus*) to 80 per cent (*Premiers Crus*) with relative prices for the grapes per kilo at harvest time. Thus a grower with a 100 per cent graded vineyard can sell his grapes for 100 per cent of the price set by the main Champagne governing body (the CIVC).

Making Champagne

The grapes are carefully picked (*l'épluchage*) and any unworthies are discarded. The great Champagne houses only use the juice from the first pressing of the grapes, known as *vin de cuvée*. When the juice becomes wine it is tasted and a *cuvée* (blend) is made incorporating the special attributes of wines from different vineyards. It was Dom Pérignon (1638–1715) a Bénédictine monk and cellar master at the Abbey of Hautvillers near Épernay who first recognized the need for compensatory blending in order to make a balanced and consistent wine. He was also the first to introduce the cork as the stopper in Champagne bottles. Previously hemp stoppers soaked in oil were used, but, of course, these were never successful in keeping the sparkle in the bottle.

Once the *cuvée* or blend is agreed a sugar solution and yeast (*liqueur de tirage*) is added. The wine is bottled and a crown cap or temporary cork, held in position by a clamp (*agrafe*), is fitted. The bottles are then taken to deep cellars (*caves*) where they are laid on their sides. A second fermentation takes place and, as the bubbles cannot escape, the CO_2 becomes chemically bonded in the wine, resulting in sparkling wine. This second fermentation is known locally as 'Prise de Mousse', capturing the sparkle. The wine is then left undisturbed to mature for two or more years.

Before the wine can be sold any sediment remaining in the bottle from the second fermentation has to be removed. The bottles are placed in a wooden frame called a *pupître* which can hold them in positions from the horizontal to the perpendicular. From time to time the cellar workman (*remueur*) twists and tilts each bottle, encouraging the sediment to slide from the body into the neck of the bottle. This

operation known as *remuage* is completed in about 90 days and the trend is towards the mechanization of this process. The bottles, neck downwards, are then placed in a tank of freezing brine so that, when the cork or crown cap is removed, the frozen sediment pops out in the form of a pellet of ice (*dégorgement à la glace*), leaving behind beautifully clear sparkling and absolutely dry wine. The small amount of wine that is lost is made up with a Champagne and sugar solution known as *liqueur d'expédition* (shipping dosage). The sweetness of the *liqueur d'expédition* depends on the country for which the wine is destined and affects the wine's style. In fact when talking about the percentage of sugar in a bottle of Champagne, this refers only to that in the *liqueur d'expédition*. The addition of sugar, as it affects the sweetness of the wine, is indicated on the label (see below).

The bottles now get their second and final corks plus metal caps which are held firmly together by a wire muzzle or cage. Each bottle is dressed with foil and a neck and/or body label.

Name	*Sugar in dosage*
Ultra brut	
Brut de Brut	
Brut absolu	None
Dosage zero	
Nature	
Brut	Up to 1%
Extra sec, extra dry	1–2%
Sec	2–4%
Demi-sec	4–6%
Demi-doux	6–8%
Doux	8% upwards

CHAMPAGNE BOTTLE NAMES AND SIZES

Name	*Metric*	*Imperial*
Quarter-bottle	20 cl	6.0 fl oz
Half-bottle	37.5 cl	12.7 fl oz
Bottle	75 cl	25.4 fl oz
Magnum (2 bottles)	1.5 litres	50.7 fl oz
Jeroboam (double magnum) (4 bottles)	3 litres	101.4 fl oz
Rehoboam (6 bottles)	4.5 litres	152.1 fl oz
Methuselah (methusalem) (8 bottles)	6 litres	202.8 fl oz
Salmanazar (12 bottles)	9 litres	304.2 fl oz
Balthazar (16 bottles)	12 litres	405.6 fl oz
Nebuchadnezzar (20 bottles)	15 litres	507.1 fl oz

Styles of Champagne

Luxury Cuvée Made by some firms in a really outstanding year and kept aside and nurtured through every stage of its development. These Champagnes are presented in beautifully elegant bottles sometimes decorated to mark a special occasion. Have your cash card ready as they are greatly expensive, but such wine is worthy of that really special occasion. Examples are:

Dom Pérignon	Tattinger Comtes de Champagne
Dom Ruinart	Pol Roger Winston Churchill
Roderer Cristal	

Vintage Champagne This is a wine from a single good year (it is however permissible to add up to 20 per cent of a wine from another specific year to assist the blend). Vintage wine will show the year on the label.

Non-vintage Champagne This is a blend of wines from different years which, for real value, is perhaps the best buy.

Pink Champagne Available as vintage or non-vintage and made by leaving the grape skins with the must until the juice becomes pink in colour. It can also be made more simply by blending together red and white wines.

Grandes Marques Champagnes Certain Houses hold this distinction because they consistently produce excellent high quality Champagne.

Ayala	G H Mumm
Bollinger	Perrier Jouët
Canard Duchêne	Piper Heidsieck
Clicquot Ponsardin	Pol Roger
Heidsieck Monopole	Pommery & Greno
Krug	Louis Roederer
Lanson	Ruinart
Moët & Chandon	Tattinger

Buyer's Own Brand (BOB) Some firms will make a Champagne for a restaurant or a chain of restaurants that will be sold under the buyer's own label.

Recently Disgorged (*Récemment Dégorgé*) (RD) These are special wines left to mature with their sediment in bottle for many years to produce a fine, full-flavoured, balanced wine. They are usually released for sale after about 8–10 years but can remain healthy for much longer.

> Note: *From 1994 other sparkling wines may not have, by law, the term Champagne method or* méthode Champenoise *ascribed to them to indicate their method of production. Instead,* méthode traditionnelle *or some other appropriate term will be substituted.*

— *Corsica* —

Corsica, the largest of France's islands produces good red and white table wine, some of which has recently received the *Vin de Corse Appellation Contrôlée* distinction. The red Patrimonio is well acclaimed, as is the white Vin de Corse Porto-Vecchio. A feature of Corsican wine is its high alcohol content. Cap Corse is a rusty red, medium sweet, wine-based aperitif that is popular on the island.

— *Jura and Savoie* —

Jura

In the districts of **Arbois**, **Côtes du Jura** and **L'Etoil** red, white and rosé (*vinsgris*) wines are made. Many are sold under the Arbois appellation. Arbois is also famous for being the birthplace of Louis Pasteur, the great French scientist who, in 1857, proved that fermentation was a physiological process, when he explained scientifically the process of vinous fermentation.

Some unique wines are also made in the Jura such as *Vin Jaune* (yellow wine), *Vin de Paille* (straw wine) and *Macvin*.

Vin Jaune This wine, made from *Sauvignon* grapes, is a wine after the style of Fino sherry. The grapes are picked late, often in December, and the cold weather induces a slow fermentation. As the wine matures in cask (sometimes for as long as eight years) a yeast mould (*flor*) develops on the surface of the wine. This imparts an austere dryness and hazelnut flavour to the pale, golden wine which has an alcohol strength of 15%. A good example, Château Chalon, is traditionally sold in a dumpy bottle called a *clavelin*.

Vin de Paille This dessert wine gets its name from the fact that the grapes are laid out on straw (*paille*) to dry and shrink them partially before pressing. Sometimes the grapes are hung from rafters during the winter to concentrate the juice. The finished wine has a flavour of quinces.

Macvin This aperitif wine, is fairly similar to white port, and is fortified with local *eau-de-vie-de-marc* and flavoured with ingredients such as coriander and cinamon. It is best served chilled and is also nice with ice.

Côtes du Jura Mousseux is the finest of the sparkling wine appellations.

Savoie

Savoie is situated between Lyon and Geneva. The area is noted for its still and sparkling white wine. Best whites are Crépy, Apremont, Seyssel, Roussette de Savoie and the sparkling wine Royal Seyssel made by Varichon and Clerc.

The reds, often made from the *Gamay* and *Pinot Noir* grapes, are best exemplified by the Cruët and Motmélian styles.

Loire —

LOIRE

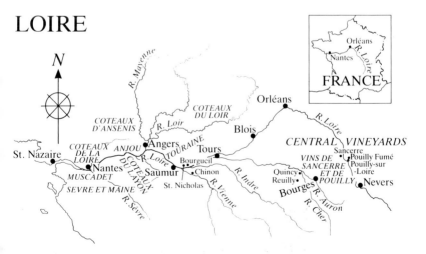

The lovely Loire valley wines are much sought after today, be they produced by a co-operative or domaine bottled from a single vineyard. Red, white, rosé and sparkling wines are produced. The reds and better rosés are made from the *Cabernet Franc* grape and the whites from *Sauvignon Blanc*, *Chinin Blanc*, *Muscadet* and *Chasselas* grapes. Although the wines may never be considered great, they are always good and present fair value in these days of upwardly mobile prices.

The Loire can be divided into four main wine growing districts: **Central Loire** (around Sancerre, Pouilly Fumé, Pouilly-sur-Loire, Quincy and Reuilly), **Touraine** (incorporating Tours, Bourgueil, Chinon and Blois), **Anjou** (including Angers and Saumur) and the **Pays Nantais** near the Atlantic coast.

The best dry whites are Muscadet, Sancerre and Pouilly Blanc Fumé.

Muscadet

Made from the *Muscadet* (*Melon de Bourgogne*) grape around the city of Nantes, this is a fresh dry wine with an attractive acidity. It is ideal as an aperitif, and goes extremely well with shellfish. The most popular style is Muscadet *sur lie*. The wine is left to mature in cask on its lees before bottling, which imparts freshness, depth of flavour and an intense bouquet.

Sancerre

While some red Sancerre from the *Pinot Noir* grape is available, the wine we have come to know as Sancerre is a delightful, dry, smoky white wine. It is produced in the Cher Department in Central Loire and can be drunk young, but it improves with bottle age.

Pouilly Blanc Fumé

A wine quite similar to Sancerre, but with a little more brightness and elegance. It gets its *Fumé* appendage not only because of its gun flint flavour, but also because of the smoky, blue, dust haze reflected by the ripe grapes over the vineyards around Pouilly-sur-Loire in the cooler autumn air.

Other Loire wines

The medium dry white wines of the Loire are Vouvray and Saumur. Vouvray, from near Tours in Touraine, is especially popular because of its versatility and genuine and agreeable nature. Saumur as a still wine is more difficult to obtain outside France. However, it is well marketed as a Champagne method sparkling wine, as is Vouvray sometimes. It is produced in the district around Saumur, in the Anjou region of the Loire. Both Vouvray and Saumur are always pleasing to taste.

Quarts de Chaume and Bonnezeaux are two of the better sweet white wines from the Côteaux du Layon area in Anjou. Made in the style of Sauternes, they improve as they mature.

The rosé wines of Anjou are very popular and those sold as Cabernet d'Anjou are of superior quality.

Of the reds, Chinon, Bourgueil, Saint-Nicolas-de-Bourgueil and Saumur Champigny are bright and fruity with a good depth of flavour. Try them slightly chilled.

— *The Midi* —

Languedoc and **Roussillon** produce everyday quaffing wines. It is estimated that up to 40 per cent of France's total production is made here. That is why the region is popularly known as the *belly of France*, because it makes huge amounts of inexpensive table wines which the French themselves are very partial to in bars, cafés and at home. While some of the wine is used as a base in the production of vermouth, the best quality (mostly red) is sold simply as Côteaux de Languedoc, Roussillon, Corbières, Fitou, Minervois and Costières du Gard.

The real stars of this wine lake are the *vins doux naturels*. Made mostly from the *Muscat* grape and sometimes from the *Grenache*, the fermentation is stopped and the sweetness retained by adding alcohol to the fermenting must. The finished wines have an alcohol content of 17%. The best examples are Grand Roussillon, Muscat de Frontignan, Muscat de Rivesaltes and the red Banyuls.

A quite outstanding dry sparkling wine, Blanquette de Limoux, is also made. Seriously considered as the best effervescent wine outside Champagne, it has elegance and fragrance and is at its most refreshing when drunk young.

The best dry white table wines are Clairette du Languedoc, Clairette de Bellegarde from Gard, and Picpoul de Pinet from Hérault. Around Montpellier, great quantities of red, white and rosé wines, known as Vins Sables du Golfe du Lion, are produced. The vines are mostly cultivated in sand dunes and in the sandy marshes close to the

Camargue. The area is especially noted for the largest single vineyard in France – Domaines des Salins du Midi – whose wines are marketed under the brand name Listel. Perhaps the most interesting of these is Listel Gris de Gris, a blush wine, made from free-run juice of the *Grenache* and *Cinsault* grapes.

— *Provence* —

Provence wines are usually sold in unusual shaped bottles with distinctive, almost lavish, labels. Those who have drunk these wines in the South of France will know they go wonderfully well with the foods and the atmosphere of the Mediterranean. Take them away from the sea and sunshine, they seem a little jaded, but perhaps it is all in the mind. Red, white and rosé wines are produced almost everywhere. The reds and rosés are from the *Cinsault*, *Cabernet Sauvignon*, *Grenache* and *Mourvèdre* grapes and the whites from the *Ugni Blanc*, *Clairette* and *Macabéo*, *Marsanne*, *Rolle*, and *Sauvignon Blanc*.

The wines

The main wine growing regions of Provence are around **Bandol**, **Bellet**, **Cassis**, **Côte de Provence** and **Palette**.

Bandol Red, white and rosé wines are produced. The reds age well, whereas the rosés are ready for drinking within a few months. The whites, though scarce (about 6 per cent of the total crop), are dry and spicy. All are fairly expensive.

Bellet Again red, white and rosé wines are made. They are rarely seen outside France as most is drunk locally in the restaurants and cafés in and around Nice.

Cassis Red, white and rosé wines are produced, although the white is predominant in quality and availability.

Côte de Provence Many Provence wines are sold under this label as well as under Aix en Provence. The rosés seem to be the best, the reds are improving, but the whites lack zip and zing.

Palette This area, near Aix-en-Provence, makes good red wines and decent white and rosé wines, though not in considerable quantities.

— *Rhône* —

The Rhône vineyards stretch from Lyon to Avignon producing red, white and rosé wines of class. The reds are the real heavyweights: big, masculine wines, strong in alcohol and flavour. They are usually made from the *Syrah* grape or from a combination of *Grenache, Cinsault,*

RHÔNE

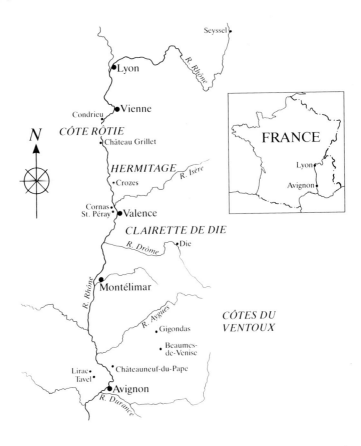

Carignan, *Mourvèdre* and others. In fact there are 13 varieties that are permitted to be used in the making of the famous Châteauneuf-du-Pape.

The outstanding reds, besides the spicy Châteauneuf, are Côte Rôtie, Hermitage, Crozes-Hermitage, Saint-Josèph, Cornas, Gigondas, Lirac, Côtes du Rhône Villages and Côtes-du-Ventoux.

The best whites are Château Grillet and Condrieu made from the *Viognier* grape. Other white grape varieties are *Rousanne*, *Marsanne*, *Clairette* and *Ugni Blanc*. Tavel is France's best rosé wine and is made predominantly from the *Grenache* grape. Saint-Péray is a clean, lively, sparkling wine made by the Champagne method, as is the better-known Clairette de Die which can also be demi-sec or doux. The great *vin doux natural* is Muscat de Beaumes-de-Venise which is fortified with grape brandy during fermentation to preserve its sweetness.

— *South-west France* —

The main wine producing regions in south-west France are **Bergerac**, **Cahors**, **Gaillac**, **Jurançon**, **Irouléguy** and **Madiran**.

Bergerac

Bergerac vineyards are located on both banks of the Dordogne where the Dordogne Valley begins. The wines are red, white and rosé, but the outstanding wine of the region is the deep golden, luscious, creamy rich Monbazillac made, like Sauternes, from grapes affected by the *Botrytis cinerea* fungus. Next best are the dry reds Bergerac, Côte de Bergerac and Pécharmant.

Cahors

This town on the River Lot is famous for its dark red wines made from *Malbec* or *Cot* grapes. In youth the wines are exceptionally robust, but usually they are allowed to age for years in cask to provide the *finesse* which has made them so popular in France. The aged wines are very good, but very expensive.

Gaillac

Gaillac is best known for its white wine made from *Mauzac* grapes. Some of the wine is *perlé* (slightly sparkling). The rosé and red wines are usually made from *Gamay* grapes and are mostly drunk locally.

Jurançon

The wine growers on the foothills of the Pyrenees, to the south and west of Pau, make mainly red and dry white wines nowadays. Some of the traditional sweet white wine called Jurançon Moelleux is still around, but it is expensive for what it is.

Irouléguy

Red, white and rosé wines are made on the western side of the Atlantic Pyrenees on the Spanish border. Although they are typical *'vins de vacances'*, the best reds are now available in Britain.

Madiran

Produces mostly red wines made almost exclusively from local *Tannat* grapes. This deep-coloured, distinctive wine is a great favourite in the Pyrenees.

GERMANY

German wines are rightly popular because they are easy to drink, can be drunk on their own as a conversation wine (Unterhaltungswein) or with a wide variety of foods. Overlooking some recent indiscretions they are consistently well made and have a reputation for reliability at all levels. Their fruity flavour, low alcohol, attractive balance of acid and sugar and reasonable price have particular appeal to new wine drinkers, which is one of the reasons why Liebfraumilch, for example, is the biggest selling white table wine in Britain today. However, it should be noted that Liebfraumilch can only ever be a *QbA* wine (see below) and can never be regarded as a classic or even fine wine as it is a blended wine from either the **Rheinpfalz, Rheinhessen, Nahe** or **Rheingau** regions. Even so, there are hundreds and thousands of people who are happy with it. To prove the point, next time you go to a 'cash wine' banquet take a look at the bottles around you.

Eighty-eight per cent of German wines are white. The rest are red and rosé (*Schillerwein*) wines. Although they are often sold under a proprietary or brand label many are identified by their region, district or vineyard and by the degree of grape ripeness at harvest time.

Sometimes, when the grapes have not fully ripened, beet or cane sugar is added to the unfermented *must* in order to raise the alcohol level during fermentation up to that of a similar wine of a good year. This practice is called *Verbesserung* (*chaptalization* in France) which 'improves' the alcohol content but not the quality. The resulting wines may not be sold as vintage wines and only the classifications of *Deutscher Tafelwein, Deutscher Landwein* and *Qualitätswein bestimmter Anbaugebiete* wines may be 'improved' in this manner.

The category *Qualitätswein mit Prädikat*, which is the top tier of German wines, may not by law be so improved. Of course with some of the lesser wines the art of the cellar master can produce a variety of styles of wine from the same basic dry wine. This is achieved by adding *Süssreserve* (sweet reserve – unfermented grape juice) to the finished wine just before bottling.

The two major categories, which are also subdivided, are *Tafelwein* (table wine) and *Qualitätswein* (quality wine).

— *Tafelwein* —

Deutscher Tafelwein (German table wine)

This is a basic German wine made only from German grown grapes. *NB* there is also a simple *Tafelwein* which is made from a combination of grapes from both Germany and the EEC.

Deutscher Landwein (German regional wine)

This is a superior *Tafelwein* wine in that it has more character and is made from grapes grown in any one of the 15 designated *Landwein* districts.

— *Qualitätswein* —

Qualitätswein bestimmter Anbaugebiete (QbA)

(Quality wine from designated regions) Wines in this category may be dry or slightly sweet. They are usually made from fully ripe grapes but, like the *Deutscher Tafelwein* and *Landwein*, the grape juice is frequently improved (*Verbesserung*) through the addition of sugar to raise the alcohol content – similar to the French '*chaptalization*'.

Qualitätswein mit Prädikat (QmP)

(Quality wine with distinction) These are completely natural wines made from fully ripe and overripe grapes. There are six levels of quality (no sugaring of the *must* is allowed):

1 *Kabinett* – gets its name from the fact that the wine is good enough to be kept in the winegrower's own cabinet (but he has let you have it instead). It is fruity, light and dry and completely genuine.

2 *Spätlese* (late harvested) – made from grapes picked about a week after the normal harvest. These riper grapes make fuller, sweeter wine.

3 *Auslese* (selective harvesting) – made from selected bunches of late harvested grapes. These wines are even sweeter, fuller and stay longer on the palate.

4 *Beerenauslese BA* (selective picked grapes) – made from individually picked grapes which, because of their overripeness, have begun to shrivel on the vine. The resulting wine is very rich (which you must also be to buy it!).

5 *Eiswein* (ice wine) – first introduced in 1842 and made from overripe grapes which have been frozen by severe frosts. The grapes are picked and pressed at a maximum of −6 °C. The grape juice is very concentrated and must have a minimum sugar content equal, at least, to a *Beerenauslese*. These wines are not made every year and can only be made when the correct conditions prevail. Although they are very sweet indeed they are not sticky or cloying as they have a welcome balance of tartness.

6 *Trockenbeerenauslese TBA* (selectively picked raisined grapes) – made from shrivelled raisin-like grapes which have been affected by *Edelfäule* (*Botrytis cinerea*) fungus. The grapes are individually picked in this prime condition and the vineyards are gone over time and time again until the harvest is over. The *Edelfäule* punctures the skins of the grapes, then, as the water content evaporates the grape sugars and acids become concentrated and the resulting grape syrup, when fermented and matured, tastes like nectar.

Deutscher Sekt, *the sparkling wine of Germany, is made in all the wine regions. Since 1986 it must be made only from German wine. The best is made by the* méthode Champenoise *(Champagne method) and is very much quality controlled. It is categorized as* Qualitätsschaumweine bA, *or* Sekt bA, *and must come from a designated region.*

*Sparkling wine is also made by the closed tank (*charmat*) method and is known locally as* Grossraumgarverwahren Schaumwein. *There are other wines with some sparkle and these are known as* Perlwein *or* Spritzig, *similar to the* pétillant *wines of France.*

Really good examples of quality Sekt *are Deinhard's Lila Imperial, Fürst von Metternich, Reichsrat von Buhl Riesling Extra Brut and Deutz & Geldermann Brut.*

— *Appellation of origin* —

Germany has 11 specified wine growing *Gebiete* (regions) (eg **Mosel-Saar-Ruwer**) which, within them, contain the 34 *Bereiche* (districts), the name of which is usually taken from the best known village of the district (eg Bernkastel). This is further broken down into 152 *Grosslagen* (sub-district or collective vineyard sites) (eg Bernkasteler Badstube) and 2,600 *Einzellagen* (individual vineyard sites) (eg Bernkasteler Doktor).

— *Label language* —

Once you understand the terminology, German wine labels are very informative and easy to understand. The label should reveal:

1 Wine category – whether it is a:
   (a) *Deutscher Tafelwein* – an ordinary wine made entirely from German grapes;
   (b) *Landwein* – which can come from any of 15 specified districts;
   (c) *Gebiet* (region) – wine made in a particular region, in eg Mosel-Saar-Ruwer;
   (d) *Bereich* – wine made in a sub-region;
   (e) *Grosslage* – wine made in a particular district or collective vineyard site;
   (f) *Gemeinde* – wine made in a village in a district;
   (g) *Einzellage* – wine made in an individual vineyard.
2 The vineyard name will be preceded by the village name in adjectival form carrying the suffix '-er', so *Bernkastel* becomes *Bernkasteler*, just as someone from, say, Dublin is known as a Dubliner.
3 Vintage year – the year the wine is made.
4 Grape variety – Riesling, Silvaner etc from which the wine was made.
5 Taste of the wine; *Trocken* (dry), *Halb Trocken* (less dry), *Diabetikerwein* (suitable for diabetics).
6 *QbA* – quality wine from designated regions.
7 *QmP* – the distinction of the wine and manner of harvesting, eg *Spätlese*, *Auslese* etc.
8 *Amtliche Prüfungsnummer* (AP number) – the official testing number indicating that the wine has passed an official chemical and

sensory test. The last two digits reveal the year the bottler made the application for his wine to be tested.

9 *Erzeugerabfüllung* – estate bottled or *Aus eigenem Lesegut* (from the producer's own vineyard).

Region

Producer's crest

Quality wine with a distinction, the distinction here being 'Spätlese'

Official proof or testing number. Last two digits reveal the year the wine was sent for testing

Village

Grape

Qualitätswein mit Prädikat

A. P. Nr. 5160087 029 89

1988 er

DÜRKHEIMER ABTSFROHNHOF

RIESLING SPÄTLESE

Erzeugerabfüllung aus dem Weingut

9,3 Vol. %

e 75 cl

Vintage year

Vineyard name

Late gathered

Estate bottled at the vineyard

Alcohol content by volume

Name and address of proprietor

Volume content of bottle

Content complies with E.E.C. bottling regulations

— *Grapes* —

Up to fifty different grapes species are grown in German vineyards, many of them new or experimental. The grape variety does not have to appear on a label but, when it does, there is a guarantee that at least 85% of the wine has been produced from the indicated grape. About 87% of the vineyards are planted with white grape varieties, the remainder being black grapes. The wine styles produced are:

- white wine
- sparkling white wine (known as *Sekt*)
- rosé wine (*Weissherbst*) made from black grapes only
- *Rotling* wine from a combination of white and black grapes resulting in such specialities as Rotgold in Baden and Schillerwein in Württemberg
- red wine.

The three noble grapes are *Riesling*, *Silvaner* and *Müller-Thurgau*:

Riesling accounts for about 21% of the crop and is the undisputed quality leader producing Germany's finest wines, elegant, well-balanced and full of flavour.

Silvaner accounts for about 8% of the crop and makes softer more gentle wines which are best drunk in their youth.

Müller-Thurgau accounts for 24% of the crop and is a cross between the *Riesling* and *Silvaner* vines. This hybrid yields good crops, is sturdy and withstands vine diseases well, producing mildly acetic, fruity wines with a pronounced flowery bouquet. It was first developed by a Professor Müller from Thurgau (Switzerland) in 1882.

Other white wine grapes are *Kerner* (a cross between the *Riesling* and the red *Trollinger* grapes), *Elbling*, *Ruländer*, *Gewürztraminer*, *Gutedel*, *Scheurebe*, *Ortega*, *Morio-Muskat* and *Bacchus*. The black grapes include *Trollinger* (originally from the Tyrol), *Portugieser* (originating in Austria) and *Spätburgunder* (Pinot Noir).

— *Wine growing areas* —

The eleven specified wine growing areas are located in south west Germany near or on the banks of rivers:

- **Baden, Hessische Bergstrasse, Mittelrhein, Rheingau, Rheinhessan** and **Rheinpfalz** are all on the banks of the Rhine
- **Franken** is on the Main
- **Württemberg** is on the banks of the Neckar
- **Mosel-Saar-Ruwer**, **Ahr** and **Nahe** are all named after their rivers.

The regions (listed alphabetically), together with examples of their better-known villages and vineyards, are given below.

Ahr

This very small region has the most northerly vineyards in Germany. The vineyards are mostly located on steep hillsides and follow the river as it flows into the Rhine, south of Bonn. The majority of the wine produced is red from the *Spätburgunder* and *Portugieser* grapes and varies from velvet-smooth to light and ordinary. The *Riesling* and *Müller-Thurgau* grapes make lively refreshing white wines. Despite their location many of the reds are on the sweet side and, like the whites, are best drunk in their own locality.

Village	Vineyard
Heimersheim	Heimersheimer Landskrone
Neuenahr	Neuenahrer Sonnenberg
Walporzheim	Walporzheimer Gärkammer

Baden

A longish, narrow stretch of vineyards and the most southerly of all the German regions. The vineyards are located between Heidelberg and Bodensee, real Black Forest country. *Riesling*, *Müller-Thurgau*, *Ruländer*, *Gutedel* and *Gewürztraminer* grapes produce a variety of white wines, some fragrant and fresh, others spicy and full of aroma. The *Spätburgunder* grapes are used to make full-bodied, smooth red wines and the popular rosé known as Weissherbst (in reality a white wine with red tinges).

Village	Vineyard
Michelfeld	Michelfelder Himmelberg
Zulzfeld	Burg Ravensburger
Durbach	Durbacher Schlossberg

Franken

The most easterly of the German regions with *Müller-Thurgau* and the *Silvaner* vines growing along the hillsides overlooking the River Main and its tributaries. The wines, which are mostly white, are also known as *Steinwein* because they are stone dry in character. They are sold in beautifully labelled, green, squat, flagon-shaped bottles known as *Bocksbeutel*.

Village	Vineyard
Castell	Casteller Kirchberg
Iphofen	Iphofener Julius-Echter-Berg
	Iphofener Kalb
Würzburg	Würzburger Innere Leiste
	Würzburger Stein

Hessische Bergstrasse

This small region lies between Darmstadt and Heidelberg on the east bank of the Rhine. The farmers mostly sell their grapes (mainly *Riesling*, *Silvaner* and *Müller-Thurgau*) to the co-operatives (*Winzergenossenschaften*) who make a pleasing white wine, generally for local consumption.

Village	Vineyard
Bensheim	Bensheimer Streichling
Heppenheim	Heppenheimer Centgericht

Mittelrhein

This beautiful region stretches south of Bonn for about 60 miles and the terraced vineyards have their hillside homes on both banks of the river Rhine. Mainly a white wine region, with the *Riesling*, *Müller-Thurgau* and the *Kerner* grapes giving lively, fruity and flavoursome wines.

Village	Vineyard
Oberwesel	Oberweseler St Martinsberg
Boppard Hamm	Bopparder Hamm Ohlenberg
Kaub	Kauber Backofen
Bacharach	Bacharacher Posten

Mosel-Saar-Ruwer

It is in this region that the classy wines begin, with the *Riesling* grape dominating. The vineyards are located on precariously steep, slatey slopes which present a severe challenge to the grape harvesters at vintage time. Since the vineyards are so widespread, the wines vary in style according to location but, at their best, they have a freshness, delicacy of flavour and lovely bouquet.

Village	Vineyard
Bernkastel-Kues	Bernkasteler Doktor
	Bernkasteler Schlossberg
Enkirch	Enkircher Steffensberg
Erden	Erdener Prälat
	Erdener Treppchen
Graach	Graacher Himmelreich
	Graacher Josephshöfer
Kasel	Kaseler Hitzlay
	Kaseler Nieschen
Ockfen	Ockfener Bockstein
	Ockfener Herrenberg
Piesport	Piesporter Goldtröpfchen
	Piesporter Gunterslay
Wehlen	Wehlener Sonnenuhr
Zell	Zeller Domherrenberg

Nahe

The vineyards are mostly found on the steep slopes along the River Nahe and its tributaries. The *Müller-Thurgau*, *Silvaner* and the *Riesling* grapes produce wines which are light and have an attractive, subtle crispness.

Village	Vineyard
Bad Kreuznach	Kreuznacher Brückes
	Kreuznacher Narrenkappe
Munster	Munsterer Dautenpflänzer
	Munsterer Pittersberg

Niederhausen	Niederhausener Hermannsberg
Schlossböckelheim	Schlossböckelheimer Felsenberg
	Schlossböckelheimer Kupfergrube

Rheingau

Regarded as the classic wine region in Germany, this region stretches from Hochheim on the River Main to Lorch close to the Mittelrhein. Most of the vineyards are situated on picturesque hillsides, with forests, castles and cloisters interspersing, making a wonderful panorama of beauty and serenity. The *Riesling* grape dominates, producing wines to match the scenery – distinctive and elegant, at once both gentle and domineering. The best reds of Germany are also made here particularly at Assmannshausen from the noble *Spätburgunder* grape.

The really great wines from this region are sold by the vineyard label and are estate bottled.

Village	*Vineyard*
Assmannshausen	Assmannshausener Höllenberg
Erbach	Erbacher Marcobrunn
	Erbacher Michelmark
Geisenheim	Geisenheimer Kläuserweg
	Geisenheimer Rothenberg
Hallgarten	Hallgartener Schönhell
Hattenheim	Hattenheimer Nussbrunnen
	Hattenheimer Wisselbrunnen
Hochheim	Hochheimer Domdechaney
	Hochheimer Kirchenstück
Johannisberg	Johannisberger Hölle
	Schloss Johannisberg
Oestrich	Oestricher Doosberg
	Schloss Reichhartshausen
Rauenthal	Rauenthaler Baiken
Rüdesheim	Rüdesheimer Berg Rottland
	Rüdesheimer Berg Schlossberg
Winkel	Schloss Vollrads

> *The wines from the village of Hochheim were favourites of Queen Victoria and became popularly known as Hocks. Indeed there is a single vineyard named after her called Hochheimer Königin Victoria Berg.*

Rheinhessen

This is the largest wine-producing region in Germany, covering an area 20 miles by 30 and lying in a valley overlooked by rolling hills, with some slopes under the vine. The area is the birthplace and the original home of Liebfraumilch – formerly only made in vineyards surrounding the Church of Our Lady (*Liebfrauenkirche*) in the town of Worms.

Although the area is huge by German standards, in terms of overall production, it still produces less wine than the Rheinpfalz. The grapes, mainly *Riesling, Müller-Thurgau* and *Silvaner*, produce the soft, mild, medium-bodied white wines for which the region is famous. The *Portugieser* and the *Spätburgunder* grapes make smooth full-bodied red wines.

Village	*Vineyard*
Bingen	Binger Scharlachberg
Oppenheim	Oppenheimer Kreuz
	Oppenheimer Sackträger
Nierstein	Niersteiner Orbel
	Niersteiner Hipping

Rheinpfalz

This region has almost 50 miles of uninterrupted vineyards and produces more wine than any other region in Germany. The main grapes used are the *Riesling, Müller-Thurgau, Silvaner, Kerner* and *Morio-Muskat*, producing the typical, rich, spicy, aromatic white wines. The *Portugieser* grape makes good smooth fruity red wines.

The region is also known as the *Palatinate* after the word *Pfalz* (a derivation of the word *palast* or *palace*, which came originally from the Latin *palatium*).

Village	Vineyard
Bad Dürkheim	Dürkheimer Fuchsmantel
	Dürkheimer Spielberg
Deidesheim	Deidesheimer Hohenmorgen
	Deidesheimer Leinhöhle
Forst	Forster Jesuitengarten
	Forster Musenhang
Kallstadt	Kallstadter Annaberg
	Kallstadter Steinacker

Württemberg

This region produces an abundance of red, white and rosé wines. Rosés are called *Schillerwein* (shimmering wines) because of their bright and breezy personality.

This region has its vineyards mainly on the banks of the River Neckar. It is Germany's largest producer of red wines, all of which are hearty and fulsome, coming from the *Spätburgunder*, *Portugieser*, *Trollinger*, *Müllerrebe* and *Lemberger* grapes. The white wines are fruity and have a distinctive earthy flavour. The commercial centre is Stuttgart and not much of the wine is allowed to leave the region. (There is a huge thirst locally!)

Village	Vineyard
Gundelsheim	Gundelsheimer Himmelreich
Maulbronn	Maulbronner Eilfingerberg
Weikersheim	Weikersheimer Schmecker

GREECE

In contrast to the wines of Cyprus, Greek wines tend to be somewhat parochial, flat and not very exciting. Best known is the famous, or infamous, Retsina which is flavoured with pine-tree resin obtained from Aleppo pines in Attica. This imparts a slight turpentine flavour which many people find pleasant to taste especially with spicy foods. It is white or pink in colour and Metaxa's Retsina is a style of note.

The dry white wines Demestica, Santa Helena, Antika and Pallini are light and pleasant, either as apéritifs or as accompaniments to light food. Demestica also appears as a red wine and the red Château Carras and the dark, fruity Naoussa are fast establishing good reputations. Two sweet wines, the deep golden Muscat of Samos and the intense red port-like Mavrodaphne, are popular dessert wines.

HUNGARY

Hungarian wines are usually named after the district in which they have been made and are marketed for export by the state monopoly called *Monimpex*. The reds are big and burly, good foils for the heavily flavoured food. The whites are also full of personality, for the same reason.

The best known of all the wines is Tokay Aszu, a luscious golden wine produced from the *Furmint* and *Hárslevelü* grapes. Heat and dampness encourage the mould *Botrytis cinerea* (noble rot) to form on the skins. These shrink, and the grape juice becomes very concentrated as the water content is reduced and the glycerine content increases. The grapes are collected in *puttonyos* (hods) – smallish wooden barrels containing about eight gallons. They are then crushed to a pasty mass and added to new must as it ferments in a standard vat. The more *puttonyos* added, the sweeter the eventual wine. The number of *puttonyos* will be shown on the label as either three, four or five *puttonyos*.

Another style of wine that used to be made is Tokay Essenz which was produced from the syrupy juice which trickled from the grapes as they lay waiting in tubs to be pressed. The resulting grape essence is not now made into wine for the commercial market, but is added to the Aszu style to enhance the product.

Many of the white table wines are sold simply as Hungarian Riesling or Balatoni Riesling but others are named: Badacsony, Mór, Somló, Pécs and Mecsek, usually with the grape appendage Riesling or Furmint.

Of the reds Egri Bikavér (Bull's Blood) is the most celebrated, but Kadarka, Vilány, Sopron and the ones labelled Hungarian Merlot are good value wines.

ISRAEL

The Carmel wine company produces over 75% of all Israeli wine, exporting mainly to the United States with some to Britain. Much is kosher made under rabbinical supervision. Many of the vineyards have now established classical grapes in order to make the more traditional table wines. These are produced in wineries, some of which were generously donated by the famous Rothschild family of France in the 1880s.

The better white wines are Carmel Hock, Château Montagne, Yarden Sauvignon Blanc and Palwin. Of the reds, Château Windsor, Gamla Cabernet Sauvignon and Adom Atic are about the best.

ITALY

The Italian wine growers once had a reputation for being careless makers of wine. Consequently they found difficulty in selling their wines abroad, except possibly to countries where immigrants from Italy had formed into sizeable communities. So, with little incentive to improve the product, the wines continued to languish, although much was used as a base for fortified and flavoured wines, spirits and liqueurs. An Act of Parliament was passed in 1963 aimed at improving the product just as the French wine laws did for French wines. However, many of the winegrowers refused to be bridled by what they considered to be petty restrictions. Even today some of the very best producers operate outside the regulation laws and can now only describe their wines as table wines.

— *The ladder of quality* —

Vino da Tavola	(table wine) – the lowest grade, plonk by another name, although some so-called table wines may be of high quality (see above)
Vino da Tavolo con Indicazione Geografica	guarantees origin not quality

ITALY

1   *VALLE D'AOSTA*
2   *PIEDMONT*
3   *LOMBARDY*
4   *TRENTINO-ALTO ADIGE*
5   *FRIULI-VENEZIA GIULIA*
6   *VENETO*
7   *LIGURIA*
8   *EMILIA ROMAGNA*
9   *TUSCANY*
10 *UMBRIA*
11 *THE MARCHES*
12 *LAZIO*
13 *ABRUZZI*
14 *CAMPANIA*
15 *MOLISE*
16 *PUGLIA*
17 *BASILICATA*
18 *CALABRIA*
19 *SICILY*
20 *SARDINIA*

Denominazione di Origine Controllata	(DOC) – indicates that the wine was made from specific grapes grown in a specific area and made and matured according to the best local custom and practice. Guarantees the origin of the wine
Denominazione di Origine Controllata e Garantita	(DOCG) – is a newish top tier classification which guarantees not only the origin but controls the type of grape, yield per hectare, minimum alcohol content, the method of viticulture and vinification. Furthermore, it guarantees that the wine has undergone a rigid chemical and sensory testing for quality and type. Only six wines have so far merited this classification: reds – Chianti, Brunello di Montalcino and Vino Nobile di Montepulciano are produced in Tuscany, while Barolo and Barbaresco are Piedmont wines; white – Albano di Romagna from Emilia-Romagna

Debate goes on about whether the laws have helped to improve the quality of Italian wine. Suffice it to say that the wines are on offer worldwide today, some horrible stuff in screw capped bottles and some splendid stuff as well.

— *Label language* —

Abboccato	slightly sweet
Amabile	semi-sweet
Annata	vintage
Asciutto	bone dry
Azienda	estate
Bianco	white
Bottiglia	bottle
Cantina Sociale (Cooperativa)	winery run by a co-operative
Casa Vinicola	wine company
Chiaretto	deep rosé
Classico	classical or best part of a particular wine area, eg Chianti Classico
Dolce	sweet
Frizzantino	slightly sparkling
Imbottigliato da	bottled by
Nero	dark red
Pradicato	control instigated by local wine growers; associated with *vino da tavola* when the wine has been produced from non-traditional grapes
Riserva	matured for a specific number of years
Riserva speciale	like *Riserva* but older
Rosato	rosé or pink wine
Rosso	red
Secco	dry
Spumante	foaming or sparkling
Spumante Classico	sparkling wine made by the Champagne method

Superiore	wines of superior quality with a good alcohol strength
Stravecchio	aged old wines
Vecchio	old
Vendemmia	harvest
Vino da Pasto	ordinary wine (*vin ordinaire*)

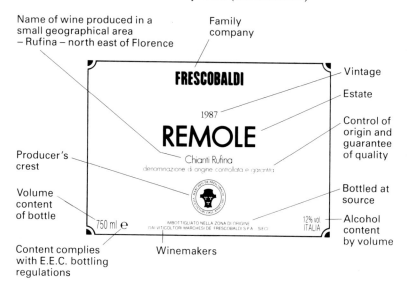

Name of wine produced in a small geographical area – Rufina – north east of Florence

Family company

Vintage

Estate

Control of origin and guarantee of quality

FRESCOBALDI

1987

REMOLE

Chianti Rufina
denominazione di origine controllata e garantita

Producer's crest

Volume content of bottle

Bottled at source

750 ml ℮

IMBOTTIGLIATO NELLA ZONA DI ORIGINE
DAI VITICOLTORI MARCHESI DE' FRESCOBALDI S.P.A. SIECI

12% vol.
ITALIA

Alcohol content by volume

Content complies with E.E.C. bottling regulations

Winemakers

— *Principal wine districts and wines* —

Piedmont

The home of vermouth which originated in Turin, the capital of
Piedmont. This district is renowned for Italy's best known sparkling
wine, Asti Spumante – *Asti* being the town where it is made, *spumante*
means foaming or sparkling. The best is made from the *Muscat* grape
and by the *Spumante Classico* method (Champagne method). Cheaper
varieties are produced by the *charmat* (closed tank) method. Two red
wines, Barolo and Barbaresco, made from *Nebbiolo* grapes are really
excellent, as is Barbera which is named after its own grape, and
Gattinara.

Tuscany

This region is renowned for its red wines, Chianti Classico, Brunello di Montalcino and Vino Nobile di Montepulciano. Chianti has traditionally been made from the *Sangiovese* grape (although others are now permitted in the blend) and is usually sold in a globular shaped bottle (*fiasco*) which is partially covered with straw – not only for appearance but, more practically, to prevent the bottles breaking when carried *en masse*. Chianti Classico wines are nowadays more usually found in Bordeaux-style bottles which facilitates their laying down for slow maturation. Watch out for the black cock neck label – this is a sign of quality. Furthermore the word *riserva* on a label implies that the wine has matured for five years in cask.

Trentino-Alto Adige

This region in the South Tyrol belonged to Austria until 1919 when it was ceded to Italy as a result of the Treaty of Saint Germain. This is mentioned because, even nowadays, the wines are marketed in both German and Italian.

The wines from **Alto Adige** are first class. The sparkling Gran Spumante is second to none in all Italy. Made by the Champagne method and from the *Riesling* and *Pinot Bianco* grapes, it is a real treat to drink and a pleasant discovery for those who think Italian spumante is nothing more than sparkling liquid sultanas. Of the whites, the Traminer Aromatico, Terlaner, Rhine Riesling and Pinot Bianco are all refreshing and good. The reds Lago di Caldaro, Santa Maddelena and Pinot Nero sell well, especially in Austria and Germany where they are very popular.

Umbria

The cathedral city of Orvieto gives its name to its most famous white wines. Orvieto Secco, which is dry, and Orvieto Abboccato or Amabile (meaning soft in the mouth) which is medium sweet. The red and white Torgiano are also important Umbrian wines.

The Marches

The dry white wines Verdicchio dei Castelli di Jesi and Verdicchio di Matelica are best here. Often sold in sexy shaped bottles, they have a pale straw colour and a slightly bitter aftertaste. The best reds are Rosso Cònero and Rosso Piceno.

Lazio

The two white wines of note here are Frascati and Est! Est!! Est!!!.
Frascati may be dry, medium or sweet. Est! Est!! Est!!! is either dry or
has a hint of sweetness. Both wines are now more famous for their
names than for the quality of their wine: The name Frascati flows and
is also easy to pronounce and remember. Est! Est!! Est!!! gets its name
from the time when the Bavarian Bishop Fugger was preparing for a
visit to Rome in 1111 AD. He sent his servent in advance to mark the
doors of various inns that he would be passing en route with the word
'est' ('it is') when the wine was good or *'non est'* when it was bad.
When the servant came to Montefiascone, he was so impressed with
the wine that he marked the door *'Est! Est!! Est!!!'*. The Bishop never
did get to Rome but stayed happily at Montefiascone until he died.

Campania

From the slopes of Mount Vesuvius come the very well known
Lacryma Christi (tears of Christ) which may be red, white or rosé – the
white is by far the best. Greco di Tufo and Fiano di Avellino are above
average whites and the reds, Ravello and Taurasi, are becoming better
known.

Sicily

This island is best known for its fortified dessert wine Marsala, a wine
that is even more popular in the kitchen than at the table. It is used in
Zabaglione as well as in veal and other dishes served *alla Marsala.* It
became popular in Britain when Liverpudlian John Woodhouse began
to develop the wine which he first exported from Marsala in 1773.

Marsala is made from two grapes, *Insolia* and *Cattarato* which
together produce a dry white wine. *Vino cotto* (unfermented grape
juice boiled in kettles to a syrup) is added, as is *vino passito*, a wine
whose sweetness has been retained by the introduction of brandy
during fermentation. The combination is put through a *solera* system
(see Sherry, page 202) to blend and mature the wine.

Today you will find Marsalas flavoured with all sorts of ingredients
such as bananas, chocolate, coffee, almonds and even egg yolks –
Marsala All'Uovo. Such additions are best avoided. The best known
shippers are Woodhouse & Co, Ingham Whittaker & Co and Florio & Co

The best Sicilian table wines are Corvo (red and white) and Regaleali
(red and white).

Sardinia

This island is especially noted for its sweet dessert wines such as Vernaccia di Sardegna, Moscato del Tempio, Malvasia di Sardegna and Anghelu Ruju. Of the table wines, the white Riviera del Corallo and the red Cannonau are best.

Veneto

The region is well known for its wine bar wines, like the dry white Soave and the popular reds, Valpolicella and Bardolino. Look out for the less well known whites, Bianco di Custoza and Verduzzo, and the sparkling Prosecco.

Emilia-Romagna

This area is famous for Albano di Romagna, the only white wine in all of Italy that has been given the *DOCG* classification. Also well known are the red, white and rosé Lambrusco Frizzante (semi-sparkling wines).

Other wine regions

Basilicata is well known for its full bodied red wine Aglianico del Vulture. In Lombardy (around Milan), **Oltrepò Pavese** makes red, white, rosé and sparkling wines. **Friuli Venezia Giulia** is well known for two red wines – Aquilea and Carso. **Liguria** is noted for Cinque Terra, dry or medium sweet white wines. **Abruzzi** makes a good red wine Montepulciano d'Abruzzo. **Molise** also has a good red called Biferno. **Puglia** has two wines of note, the red Il Falcone and the pink Rosa del Golfo. **Calabria** is best known for Greco di Bianco, a lovely, big, creamy sweet wine (Vino Passito). Italy's smallest wine region **Val d'Aosta**, produces honest wines which are almost always drunk locally.

LEBANON

Despite all the political troubles, red and white wines continue to be made in this war-torn land. The reds are much better than the whites, being made from a blend of *Cabernet Sauvignon, Syrah* and *Cinsault* grapes. Best examples are: Château Musar and Cuvée Musar which come from a single 140 hectare estate owned by the Hochar family. These wines have good ageing qualities.

LUXEMBOURG

Luxembourg makes quite a number of thinish white wines from grapes such as *Riesling*, *Elbling*, *Gewürztraminer* and *Sylvaner*. There is also a slightly sparkling wine called Edelperl.

MALTA

Cultivation of the vine is not easy in Malta where the climate can vary between torrential rain and scorching sunshine. The result is ordinary, even harsh wine.

All types of wine are made, with the Altar wine, the wine of the church, perhaps the best of all. There are pleasant dessert wines made from the *Muscat* grapes, and the winery Marsovin produces palatable red, white and rosé table wines. Others of note are Verdala Rosé, Lachryma Vitas (red and white), Coleiro (red and white) and the Farmers' Wine Co-operative which also produces red and white table wines.

MEXICO

The home of Tequila is now getting a good reputation for wine. The best is made from varietal grapes: *Cabernet Sauvignon*, *Barbera*, *Malbec*, *Merlot*, *Trebbiano*, *Grenache*, and *Zinfandel* for red, and *Chenin Blanc*, *Sauvignon Blanc*, *Riesling* and *Chardonnay* for whites.

Well-known producers are: Bodegas de Santo Tómas, Casa Madero, Casa Martell, Casa Pedro Domecq, Cavas de San Juan, Antonio Fernandez y Cia, Marqués de Aguayo and the largest winery of all, Vinicola de Aguascalientes.

MOROCCO

When France established control of Morocco in 1912 much effort was made to upgrade the existing vineyards and to cultivate new ones. Most of the resulting wine was exported to France and the French colonies. Stricter French wine legislation now makes it impossible to import bulk Moroccan wine and sell it or export it under a French label, so the business is not what it was.

Plenty of red and white wines are made, with **Fez** and **Meknes** the chief wine producing centres. The red wines are considered best, and improve with bottle age. Tarik and Chante Bled, full-bodied and well balanced are typical examples.

A unique wine of special interest is Gris de Boulaouane. It is a blush wine produced by the bleeding method. The grapes are suspended on sheets of white linen where they are self-pressed by their own weight. The juice slowly drips through the linen into containers and the resulting wine is aged in bottle, not in cask.

NEW ZEALAND

There has been a distinct improvement in New Zealand wines over the past ten years. Previously the country was associated with very ordinary fortified wines but today some excellent table wines are produced. The **North Island**, particularly **Auckland** and its surrounds, was the traditional homeland for the vine, but more significant vineyards are now to be found on the East Coast of the North Island in regions like **Gisborne** and **Hawkes Bay** and on the north edge of the South Island at **Marlborough**.

The first vines were planted in 1819 but the *phylloxera* bug later took its toll. It was only in the 1970s and onwards that New Zealand came to grips with winemaking. Many vineyards previously planted with hybrid vines, were dug up and replanted with classic *Vitis vinifera* styles. Initially the *Müller-Thurgau* vine was the standard bearer, producing in abundance good dry and medium dry wines. With its twin virtues of high yield and early ripening, the success of the *Müller-Thurgau* encouraged the planting of finer varietals such as *Sauvignon Blanc, Chardonnay, Rhine Riesling, Gewürztraminer, Sémillion, Pinot Blanc* and *Chenin Blanc*. Although the *Müller-Thurgau* still accounts for about 30% of the total output, it is the *Sauvignon Blanc* and the *Chardonnay* that are enhancing New Zealand's evergrowing reputation for fine white wine. Traditional and high-tech methods have, so far, failed to yield red wines of comparative quality, but experimentation with the *Cabernet Sauvignon, Pinot Noir, Merlot* and *Pinotage* vines continues.

The principal wineries in New Zealand are: Babich, Cloudy Bay, Collard Brothers, Cooks, Coopers Creek, Delegat's Vineyard, Hunter's Wines, Mission Vineyards, Montana Wines, Nobilo's, Penfolds Wines, Selak Wines, Te Mata Estate and Villa Maria Estate.

PORTUGAL

For such a small country, Portugal produces a wonderful array of wines from table wines to the two classics, port and Madeira. The table wines, although not nearly as renowned as the fortified ones, reach a good average standard.

It was the Methuen Treaty of 1703 that brought Portuguese wine to the attention of the British wine drinkers. The terms of the Treaty between England and Portugal gave preferential treatment to the wines of Portugal over the traditional wine source suppliers, France and Germany. The wines then were extremely good value and have remained so to this day. A great variety of grapes are used in the production of these wines some which are listed below.

Red wine grapes		*White wine grapes*	
Agua Santa	Ramisco	Alvarinho	Malvasia
Alvarelhão	Tinta Pinheira	Arinto	Moscatel
Bastardo	Touriga	Galego Dourado	Rabigato

— *Wine regions* —

The main wine regions (excluding Madeira) are **Bairrada**, **Beiras**, **Bucelas**, **Carcavelos**, **Colares**, **Dão**, **Estramadura**, **Minho**, **Setúbal**, and **Trás-os-Montes**. These regions are considered below.

Bairrada

This region in the west of Portugal is known for good quality red wines which are rich in tannin when young but which, with patient ageing, become mellow and soft. White wine of average quality is made for local consumption as are some agreeable sparkling wines. The best of the latter comes from the Quinta do Ribeirinho and is made by the *méthode Champenoise*.

Beiras

Situated in the far north of Portugal, this area produces some rosé wines around the town of Pinhel. The region is also noted for an excellent red wine Buçaco (Bussaco) and a good *méthode Champenoise* sparkler called Raposeira. The white wines tend to be slightly acidic.

Bucelas

Located 16 miles north of Lisbon, this small region produces white wines from the classic *Arinto* grape. However, despite the noble grape, the wines are very ordinary.

Carcavelos

This is a small vine growing area between Lisbon and Estoril. It is especially noted for its fortified almond-flavoured wine which is made at the Quinta do Barão.

Colares

These vineyards are situated by the sea about 14 miles north-west of Lisbon. Because of the nature of the soil, the *phylloxera* aphid could not penetrate the great carpet of sand and consequently the vineyards escaped the terrible scourge. The *Ramisco* grape is used to great effect in the making of Colares, producing what many consider to be the premier red wine in Portugal. In youth the wine is very astringent but age matures and mellows it out to a silky smoothness. Some undistinguished white wine is also made.

Dão

Situated in the centre of Portugal, this area is famous for its full-bodied, strong, earthy red wines. They are mostly blended and have an agreeable smoothness because of their unusually rich glycerine content. The white Dão is greatly respected and appreciated locally but not much reaches the export market.

Estramadura

This most prolific region in all Portugal is located 70 miles north-west of Lisbon. The red and white wines are basic table wines meant for everyday drinking. Some of the whites can have a slight effervescence.

Minho

This region produces one of the most distinctive wines in Portugal. Vinho Verde (green wine) is made close to the Spanish border but it is not a green wine as such. The 'green' refers to the youth and personality of the wine (which come from the use of some underripe grapes) and not the colour, which is either red or white. About three times more red than white is made, but curiously it is the white variety that is mainly exported.

The wines are made from grapes that are grown high up on pergolas. The grapes are picked early when they are slightly underripe. Because of this special method of vine cultivation the grapes get less reflected sunshine, resulting in grapes with proportionally less sugar but a higher malic acid content. Once bottled, the malic acid is broken down by naturally occurring bacteria. This evolution creates a malolactic fermentation which does not increase the alcohol content but produces a slight and agreeable effervescence. Vinho Verde should be drunk when it is young and vigorous. It does not require ageing.

Setúbal

Situated to the south-east of Lisbon, this area is famous for the fortified, intensely sweet, amber coloured Moscatel de Setúbal. The wine is usually aged in cask from six to twenty-five years, although younger and older examples are available. As it ages it develops an attractive honey flavour.

Trás-os-Montes

This region's name is synonymous with the internationally famous Mateus Rosé. The wine is so popular that the grape harvest of a great many vineyards in this rugged, mountainous region of the upper Douro River is given over to the production of the pink, *pétillant* wine. The wine is beautifully labelled and presented in flagon-shaped bottles which, however attractive, are difficult to store. The bottles, once prized as decorative lamp shade bases, seem to have lost their appeal in this respect.

— *Label language* —

Adega	winery
Colheita	vintage
Engarrafado	bottled by
Garrafeira	tells you the wine has matured in cask and bottle for some time – at least one year for white wine and two years for red wine

Quinta	estate
Região Demarcada	the wine comes from a legally demarcated region
Reserva	a quality aged wine
Selo de Origem	seal of origin
Denominacao de Origem	guarantee of origin and quality, similar to *DOC* in Italy
Vinho espumante	a sparkling wine
Vinho generoso	a strong, dessert wine

Region, name and style of wine

Comes from a legally demarcated area in Portugal

VINHO VERDE

REGIÃO DEMARCADA
BRANCO — White

QUINTA DE ANCEDE — Name of the property, estate or farm

Bottled on the estate — ENGARRAFADO NA ORIGEM

JOÃO BARBOSA DE MACEDO — Owner/producer

Village name — PROZELO

AMARES

Town in the Minho — PRODUCE OF PORTUGAL — 750 ml — Volume content of bottle

— *Port* —

This great after dinner drink is made from a combination of grapes grown in the upper Douro Valley, in northern Portugal. At a precise moment during fermentation the wine is transferred to vats where sufficient local high strength brandy (*aguardente*) is added to halt the fermentation. This not only fortifies the wine but ensures the retention of some of the grape sugar, giving a finished product that is soft, sweet and smooth.

Port is a blended product of wines from different *quintas* (farms or estates) and, although made in the Douro, the wine is matured in *lodges* (cellars) in Vila Nova de Gaia across the river from Oporto.

There are two main styles of port, each with its own subcategories:

Wood port

This is port that has spent all its life maturing in cask and is ready for drinking once bottled. Examples are white, ruby and tawny port.

White port Made from white grapes, it is usually sweet but sometimes an apéritif style is produced. An example of the latter is Cockburn's Dry Tang which is deliciously refreshing either chilled or with ice and a slice of lemon.

Ruby port This type of port is kept in cask for about five years until it becomes ruby in colour. It is a basic port often drunk with lemonade and other mixers.

Tawny port Tawny port is kept in cask for up to ten years (more for an aged tawny) or until it fades to a tawny colour. Although, like ruby, it is a blend of different years, it can be a mellow, real quality wine.

Bottle port

This is port that has matured mainly, or at least some time, in bottle. Examples include crusted, late-bottled vintage (LBV), vintage character and vintage.

Crusted port This is a blend of vintage quality ports kept in cask for up to four years. When bottled it throws sediment or crust as it matures – hence the name.

Late-Bottled Vintage (LBV) LBV is a vintage-style port, matured in wood for about six years, and then bottled where it will continue to improve.

Vintage character port This port is a blended good quality wine of different years. Rather similar in style to LBV but not as classy, it is wooded for about four years and is ready to drink soon after being bottled.

Vintage port Vintage port is made from grapes grown in an outstanding year, producing the finest port available. Matured in wood for two, sometimes three years, it is kept in bottles for up to 20 years or more so that it matures slowly. It is always binned and handled with the whitewash splash or label uppermost so as not to disturb the sediment.

Like crusted port, the vintage variety should always be decanted (see page 122). When dining the decanter is traditionally passed to the person on one's left (the way of the sun or, more romantically, so it is given from the heart).

Brand names

The following brand names are generally associated with port:

Cockburn's	Ferreira	Quinta do Noval
Croft	Fonseca	Sandeman
Delaforce	Graham's	Taylor's
Dow's	Gonzales Byass	Warre's

— *Madeira* —

When the island was first discovered in 1418 it was covered with trees and was given the name Madeira (meaning 'wood'). Legend has it that it took seven years to burn the trees to ashes. The ash enriched the soil, providing a suitable home for the vine and sugar cane. The wine Madeira was first fortified in the eighteenth century to strengthen it for long sea voyages. Casks of wine were used as ballast as ships journeyed from Funchal (the capital) to the Portuguese colonies, South America and the Far East.

It was soon discovered that the combination of heat and agitation as the wine was carried to and from the tropics improved the taste of the wine. Later the *estufa* system was introduced to give similar results.

The estufa (stove or heated room) has concrete vats in which the wine is gradually heated to an agreed temperature limit, usually not more than 50 °C (120 °F) and then slowly brought back to normal temperature. This 'cooking' of the wine, which can take four to five months, imparts colour and a special flavour unique to Madeira. The wine is blended and matured by the *solera* system (see page 202).

Styles of Madeira

There are four main styles of Madeira, each of which is named after the principal grape used:

Sercial This is the driest Madeira produced. It makes a good apéritif chilled or with ice.

Verdelho A sweeter richer wine that is also used in the kitchen to flavour soups and sauces.

Bual Pronounced 'boal', this is a full, deep golden wine that is sweet without being cloying on the palate.

Malmsey A luscious dessert wine that is darkish brown in colour. It is initially very sweet on the palate but, like all Madeiras, it leaves you with a 'dry goodbye' which is a quite delicious contrast.

Brand names

The following are the brand names frequently associated with Madeira wine:

Blandy Brothers	Leacock	Rutherford & Miles
Cossart Gordon	Lomelino	

ROMANIA

Romanian wines have made little impact on world markets even though they are very reasonably priced. Those to look out for are the late harvested Gewürztraminer and the Rosé Edelbeerenlese made from botrytized grapes (grapes attacked by *Botrytis cinerea*). The light red Valea Lunga, the spicy red Nicoresti, the fulsome Tohani and Valea Calugareasca (Valley of the Monks) are all good with food. Last but not least is Cotnari, the lush white dessert wine made after the style of Tokay.

SOUTH AFRICA

The first vines were planted in South Africa in 1654 but it was not until the nineteenth century that their wines became popular worldwide. Today alas, political pressure has hit the wine industry hard. Although the wines are available in Britain, they are seen less and less frequently on wine lists, thus preventing those so-minded from enjoying their undoubtable quality – and the wines are such good value too. Most bottles show a coloured seal of origin known locally as the 'bus ticket': *blue* guarantees the location of production; *red* guarantees that at least 75% of the wine was made in the year indicated on the label; *green* certifies that at least 75% of the wine was made from the indicated grape; and *gold* suggests a wine of superior quality.

In 1918 the *Ko-öperatiewe Wijnbouwers Vereniging (KWV)* – the Co-operative Winegrowers Association – was formed to organize and supervise the production methods and marketing of the industry. They have done a good job and about 90% of the wine exported comes from members of this association. Incidentally the *KWV* cellars at Paarl cover 25 acres and are capable of holding 30 million gallons of wine.

— *White wine* —

The best white wines come from the areas around **Stellenbosch, Paarl** and **Tulbagh**, with *Riesling, Sauvignon Blanc, Clairette Blanche* and *Steen* the favoured grapes. Owing to the hot climate the wines undergo a slow, temperature controlled fermentation to preserve quality.

South African white wines are meant to be drunk young when they are lively and refreshing. Good examples are:

Fleur du Cap
   Sauvignon Blanc
KWV Chenin Blanc
Groot Constantia
   Gewürztraminer

Twee Jongegezellen
Theuniskraal Riesling
Zonnebloem Noble Late Harvest
   Superior

— *Red wine* —

The reds are mainly produced in **Constantia, Durbanville, Paarl** and **Stellenbosch**, where *Cabernet Sauvignon, Shiraz, Gamay, Pinot Noir* and *Pinotage (Pinot Noir* and *Cinsault)* are the most used grapes. South African reds may be light- or full-bodied. Look out for:

Zonnebloem Cabernet Sauvignon
Culemborg Pinotage
Nederburg Cabernet

KWV Roodeberg
Château Libertas
Backsberg Estate

— Other drinks —

South Africa makes the best sherry outside Spain, particularly the *fino* variety which is really outstanding. *Flor*, a thin layer of yeast fungus forms naturally on these wines, which enhances the flavour as they mature in cask. The best finos come from **Stellenbosch** and **Paarl** and are matured by the *solera* system.

The sweeter sherries are made in **Worcester**, **Robertson**, **Montagu** and **Bonnievale**, areas which also produce dessert wines such as the Muscatels and many styles of port.

The Oude Meester Company make a fine brandy called Oude Meester (Old Master) and the best known liqueur from South Africa is the tangerine flavoured Van der Hum.

SPAIN

Spain has the largest area of vineyards in the world yet it is only the world's third largest producer. Closer planting of vines and improved methods of viticulture will, eventually, improve volume, but quality is already there as the country produces excellent wine in a wide variety

of styles. The production is controlled by 28 denominations of origin –
Denominación de Origen (DO) – which regulate viticulture and
vinification standards and set guidelines for marketing, promotions
and sales. Of all the wines produced sherry is the most famous.

— *Sherry* —

This fortified wine is produced in southern Spain in the Provence of
Cadiz. The vineyards are situated around three towns, **Jerez de la
Frontera, Sanlúcar de Barrameda** and **Puerto de Santa Maria**. While
some of the vines are grown in clay (*barros*) and sand (*arenas*) the most
hospitable soil for the two grapes used – *Listan* (*Palomino*) and *Pedro
Ximénez* (*PX*) – is chalk (*albariza*). About 85% of the vineyards are
under the *Palomino* vine which produces a basic dry white wine, while
the *PX* grapes are used to enrich, with sweetness, the heavier styles of
sherry destined for export. As all sherries are naturally dry, any
sweetness has to be added.

In winter the new wine is put into small nursery casks or *criaderas*
(cradles) to see how it is going to develop. It is carefully monitored and
in spring a yeast-like fungus (*Saccharomyces beticus*), also known as
flor (flower), will develop on the surface of some of the wines. This

spontaneous occurrence is an indication that the wine will develop into
the highly prized, light, delicate wine known as *fino*. Absence of *flor*
shows the wine is destined to become a full-bodied *oloroso*. Of course
there are variations of these two major classifications as we shall see.

The *finos* are now fortified up to 15.5% volume of alcohol by the
addition of local high strength brandy; the olorosos are topped up to
18% by volume. Both types usually get a second fortification before

sale – bottled sherries have an alcoholic strength of between 16 and 21% by volume. The sweeter styles are the more heavily fortified.

Before bottling, all sherries must go through the unique maturing and blending system of the *solera*. The *solera* is a series of casks placed one on top of the other five or six scales high. When wine is required for sale it is drawn from the bottom scale or casks – only one-third is allowed to be drawn off each year. The void is replenished by wine from the scale immediately above. This continues upwards, with the more mature wines being continually refreshed by younger wines – the youngest of all being in the top scale. Finally the wine for sale is blended and adjusted, if necessary, for colour, sweetness and alcohol content. The skill of the blender (the sauce chef of the wine trade) ensures consistency of product.

Some old unblended sherries known as Almacenista are also made. These dry, exceptional wines are of supreme distinction and are used primarily as blenders to give character and uplift to the more mundane styles. However some firms are now selling them as a special brand in their own right.

Styles of sherry

Fino A fragrant, delicate, pale and extremely dry sherry. It has a powerful bouquet and pleasing flavour.

Manzanilla A style of *fino* made from grapes grown by the sea at Sanlúcar de Barrameda and matured in local *bodegas* (cellars). Proximity to the ocean breeze and the unique thickness of the *flor* imparts a salty and attractive bitter tang to this wine.

Amontillado Slightly fuller and deeper in colour than *fino*, it is a member of the *fino* family and has a pronounced nutty flavour. This wine gets its name from its similarity in style to Montilla wine produced near Cordoba 120 miles away.

Palo Cortado Somewhere in style between an *amontillado* and an *oloroso*. It is very popular in Spain and difficult to find elsewhere.

Oloroso Rich and deep coloured with a generous, full flavour, it is sweetened by *Pedro Ximénez* grape concentrate known as *dulce*. The name *oloroso* encompasses all the cream, brown and East India Sherries on the market. However many of the pale cream sherries marketed are not true *olorosos* but are sweetened *finos*.

Service

Fino, *manzanilla* and *amontillado* sherries are best served chilled or on the rocks. Others styles may be served straight from the bottle, ideally into a *copita* – the traditional sherry glass.

Using a 3 out measure (⅓ of a gill) (1 gill = ¼ pint), 16 measures can be obtained from a normal sized bottle (75 centilitres = 26 ⅔ fluid ounces).

Sherry shippers

The best known sherry shippers (producers) are:

Gonzalez Byas	Garvey	Harveys	Sandeman
Pedro Domecq	Williams & Humbert	La Riva	
Croft	Duff Gordon	Osborne	

— *Table wines* —

Nearly all the best wines of Spain come under the *Denominación de Origen (DO)* laws, which are similar to the *AOC* of France and *DOC* of Italy. Whereas sherry is produced in the south, most of the finer table wines are produced in the north in regions such as **Rioja** and **Catalonia**.

Rioja

Tempranillo, Garnacha, Graciano and *Mazuelo* grapes produce Spain's famous red wines. The Rio Oja, a tributory of the Ebro, gives its name to the locality which is divided into three regions: **Rioja Alta** (the best Rioja area), **Alavesa** (the next best) and **Rioja Baja**.

The wines are big, soft, rich and mellow with a distinctive oaky flavour derived from being matured in small 225 litre oak casks for up to six years. Best examples are Marqués de Murrieta, Marqués de Riscal, Marqués de Cáceres, Vina Tondonia, La Rioja Alta, of which the style Viña Ardanza Reserva 904 is quite outstanding, and CVNE (Compania Vinicola del Norte de Espana).

Some good white Riojas from *Malvasia* and *Viura* grapes are also made, such as Marqués de Cáceres, Marques de Murrieta, Viña Soledad and CVNE.

Catalonia

Penedés Excellent sparkling and still white and red wines are made here. The sparkling variety is sold under the generic name of *cava* (cellar). Made by the *méthode Champenoise*, it is widely available under proprietory brand names such as Gran Codorniu, Freixenet Cordon Negro, Segura Viudas and Castellblanch Brut Zero.

The white wines of Penedés are a revelation especially Torres Gran Viña Sol Green Label made from *Parellada* and *Sauvignon Blanc* grapes. **Torres** also makes Viña Esmeralda, a medium sweet wine made from *Muscat* and *Gewürztraminer* grapes. The dry, crisp Jean León Chardonnay and Marqués de Allela are produced near the suburbs of Barcelona. Of the red wines, Torres again produces the pedigree wines Gran Coronas (*Cabernet Sauvignon* and *Tempranillo*) and Gran Coronas Black Label (*Cabernet Sauvignon* and *Cabernet Franc*) – the latter is a really wonderful wine. Jean León's Cabernet Sauvignon is also excellent.

Ribera del Duero Ribera del Duero produces, in small quantities, the great classic red, Vega Sicilia Unico Reserva – a wine that ages well up to 30 years and is considered one of the finest reds in the world. It is made from the *Cabernet Sauvignon, Malbec* and *Merlot* grapes and is frightfully expensive. Tinto Valbuena is another fine, slightly less expensive, wine produced in the same *bodega* near Valladolid. Look out for Bodega Alejandro Fernandez Tinta Pesquera and Bodega Hermanos Perez Viña Pedrosa which offer much better value.

Other wine producing regions

Navarra The region of Navarra is located on the border of Rioja and spills over into Rioja Baja. The red, white and rosé wines from this area get heavy media promotion as alternatives to Rioja, but they have a long way to go since, at the moment, they are far too robust by comparison. Watch for the names Bodegas Villafranca Monte Ory Reserva, Julián Chivite Gran Feudo Tinto, Agro Navarra Camponuevo Tinto, Señorío de Sarría Viña Ecoyen Tinto and Viña del Perdon Tinto.

Andalucia This region produces a lovely fortified dessert wine, Málaga, once known as 'Mountain', that is rarely seen nowadays outside Spain. Made from the *Pedro Ximénez* and *Muscat* grapes, it undergoes the blending and maturing process of the *solera*. The best style is Lágrima ('tear' – as in 'weep') made from grapes which are self-pressed by their own weight. Bodegas Barcelo and Scholtz Hermanos make quality Málaga.

La Mancha The Midi of Spain produces a vast wine lake of table wines – often the base for Sangria and headaches! Valdepeñas (Valley of Stones) produces sound red and white wines which are used primarily as carafe wine in bars, *bodegas* and restaurants in and around Madrid.

Alicante This region is best known for its rosé and for the extremely dark red Vino de Doble Pasta – made by adding a double quantity of grape skins which deeply darkens the colour during fermentation.

Ampurdán (Costa Brava) This area is best known for its sparkling wine, Perelada, which is made by the *méthode Champenoise*.

— *Label language* —

Abocado	medium sweet
Año	year
Blanco	white
Bodega	winery
Cava	cellar or generic name for sparkling wine made by the *méthode Champenoise*
Clarete	light red table wine
Cosecha	vintage

Denominación de Origen (DO)	a quality guarantee by the wine's governing body
Consejo Regulado	an assurance that the wine comes from the locality stated on the label
Embotellado por	bottled by
Espumoso	sparkling
Gran Reserva	the highest grade for a quality wine which has spent two years in cask and three years in bottle
Reserva	the next best grade – the wine has matured one year in case and two years in bottle
Rosado	rosé, pink
Seco	dry
Semi-seco	medium dry
Sin crianza	without wood age
Tinto	red
Vendimia	vintage harvest
Viña	vineyard
Vino de mesa	table wine

Producer's crest

De luxe brand. Wine made from grapes grown in Rioja Alta

Top quality style. This red wine must be aged for a minimum of 2 years in cask and 3 years in bottle before being released for sale

Controlled–quality wine region

Town

Alcohol content by volume

Volume content of bottle

IMPERIAL
GRAN RESERVA
Rioja
Denominación de origen

Compañia Vinícola del Norte de España, s.a.

13% Vol.   HARO·LA RIOJA

75 cl. e   R.E N° 44 LO   PRODUCE OF SPAIN

1981

Name of wine and region

Name of wine company/ producer

Stamp of regulating body

Content complies with E.E.C. bottling regulations

Registration number of the producer given by the regulating body

Vintage

SWITZERLAND

Without a doubt, the quality of Swiss wine is consistently good but the wines are expensive to buy and fairly difficult to obtain. The country is made up of 25 *cantons* (districts), most of which produce wine, with the French speaking *cantons* producing the best of all.

— *Valais* —

The **Valais** wine-growing area extends along the entire valley of the Rhône from Viège to Martigny. The region produces good red and white wines – the red Dôle is well considered. Made from *Gamay* and *Pinot Noir* grapes, it has an affinity to the fulsome Burgundy reds. Petite Dôle is also an interesting red wine made solely from *Pinot Noir* grapes. Of the whites, Fendant made from the *Chasselas* grapes is best, but there are other good styles such as the Johannisberg.

Vin du Glacier is a white wine made in the **Anniviers Valley** and then taken to caves in the mountains near the glaciers to mature in larchwood casks for up to 15 years.

Malvoisie is an estate bottled, straw coloured, dessert wine made from late gathered noble rot grapes.

— *The Vaud* —

Vaud vineyards are located along the shores of Lake Geneva (Lac Léman) and include three smaller districts: **La Côte**, **Lavaux** and **Chablais**. The area is generally known for its fine white wines made from the *Chasselas* (*Dorin* locally) grape. Prime examples are Dézaley, Mont-sur-Rolle, Saint Saphorin and Aigle Clos de Murailles. The latter is one of the finest and most expensive of all Swiss wines.

— *Neuchâtel* —

Neuchâtel produces light fragrant white wines mostly from the *Chasselas* grape. Some are *pétillant* and produce the famous Neuchâtel Star when poured from a height into the glass. There is also a red *Pinot Noir* wine called Cortaillod and a pink one called Oeil de Perdrix (partridge eye) made from the same grape.

TURKEY

Much of the grape production in Turkey is used for table grapes and sultanas, but some reasonable wine is also made. The majority of the wine is red, good examples being Villa Doluca, Hosbag, Buzbag and Trakya Kirmisi. The dry Trakya made from the *Sémillon* grape is the best of the whites.

UNION OF SOVIET SOCIALIST REPUBLICS (USSR)

The Soviet Union makes huge quantities of very ordinary wines which are generally on the sweet side to suit the Soviet palate. For this reason the wines have not been taken on abroad, but there are rumblings that standards are improving and that the government is encouraging people to drink wine in preference to vodka – but old habits die hard!

The principal regions are **Moldavia**, **Crimea**, **Georgia** and **Armenia**.

— *Moldavia* —

This region produces good reds, including Negru de Purkar and Kabernet, and the whites Aligoté and Fetjaska.

— *Crimea* —

An abundance of dessert-style wine after the taste of sherry, port, Madeira, muscatel and tokay are produced here. Massandra is the centre for these. Good sparkling wines, Kaffia and Krim, are made by the *méthode Champenoise*.

— *Georgia* —

This region exports its best wines – the reds Mukuzani and Nepareuli, the straw coloured whites Tsinandali and Gurdzhaani, and the sparkling Krosnador.

— *Armenia* —

Port, sherry and Madeira style dessert wines are produced here. The red Norashen, the white Echmiadzin and the pink Pamid are the best table wines. The sparklers are Perla (white) and Iskra (red) both made by the *méthode cuve close*.

UNITED STATES OF AMERICA (USA)

Although the vine can prosper in almost all of America's states, the traditional ties are with the southern seaboard – the coastal plain between Virginia and Florida. Wines are also associated with Washington, Missouri, Ohio, Maryland and Michigan. However, the principal wine producing areas are in the states of California, New York and Oregon.

— *California* —

With a climate ideal for cultivating the vine, California produces 90% of American wine in all sorts of styles from jug wines, known here as *carafe* wines, to the 'designer' wines produced in the 'boutique wineries' of the **Napa Valley**, **Sonoma**, **Santa Clara** and **Mendocino** areas.

The jug wines, albeit attractively presented in carafe style containers *à la Paul Masson*, were the first introduction for many to American wine. Wine drinkers are now looking for a more sophisticated, elegant product, which is being met by winemakers, many of whom are graduates of the Wine College of the University of California at Davis. Concentration is now on classic *varietals* (wine made from a single grape variety) and on technical innovations which will help nature make good wines even better. The Americans believe they now have the ability to 'design' a wine to suit every market. Grapes, of course, will always be needed – the two glamour grapes being *Chardonnay* for white wine and *Cabernet Sauvignon* for red.

Besides the majestic *Chardonnay*, other varietals for making white wine are *Chenin Blanc, French Colombard, Gewürztraminer, Pinot Blanc, Sauvignon Blanc, Sémillon, Johannisberg Riesling* and *Muscat*. The red wine grapes, headed by the *Cabernet Sauvignon*, are *Merlot, Pinot Noir, Syrah* and *Zinfandel*.

Varietal grapes and wines

The main grapes used, together with some of the best styles of varietal wines produced in California, are given below. Names in brackets refer to specific wine growing areas, eg the Napa and Sonoma valleys.

White grapes

Chardonnay

Chalone (Monterey)
Robert Mondavi Winery (Napa)
Château Montelana (Napa)
Château St Jean (Sonoma)
Stony Hill (Napa)

Mount Eden (Santa Clara)
Trefathen (Napa)
Clos du Bois (Sonoma)
Mantanzas Creek (Sonoma)
Buena Vista (Sonoma)

These outstanding producers of Chardonnay make elegant wines which have a very fine balance with apple fruit, vanilla, oak and slightly smoky flavours.

Sauvignon Blanc

Robert Mondavi Winery (Napa)
Kalin (Marin)

Matanzas Creek (Sonoma)
Lyeth (Sonoma)

These are light, fresh wines with a good acidity balance and a distinct flintiness of flavour.

Chenin Blanc

Dry Creek (Sonoma)
Preston (Sonoma)

Hacienda (Sonoma)
Chappellet (Napa)

Light bodied and delicate with a good degree of acidity, these wines still have lots of room for improvement.

French Colombard

Parducci (Mendocino)   Chalone (Monterey)

These are light, aromatic wines with an attractive crisp flavour. They are associated with good value, unpretentious, jug wines.

Gewürztraminer

Evensen (Napa)
Alexander Valley (Sonoma)

Joseph Phelps (Napa)
E & J Gallo (Central)

These producers make dry, slightly sweet or sweet (from late-gathered grapes) wines. The wines are often quite good but bear no resemblance to the classic Alsace wine of the same name.

Pinot Blanc

Chalone (Monterey)	Congress Springs (Santa Clara)
Bonny Doon (Santa Cruz)	Mirassou (Santa Clara)

After the style of Chardonnay but not as generous, these wines are crisper and firmer to the palate.

Sémillon

Kalin (Marin)   Robert Mondavi Winery (Napa)

Often blended with the *Sauvignon Blanc*, the wines produced have good body and a rich, creamy flavour.

Johannisberg Riesling (dry and sweet)

Clos du Bois (Sonoma)
Jekel (Monterey)   make the dry, tangy
Kendall Jackson (Lake)   varieties
Château St Jean (Sonoma)

Joseph Phelps (Napa)
Raymond (Napa)
Firestone (Santa Barbara)   make the lusciously
Kenwood (Sonoma)   sweet varieties
Château St Jean (Sonoma)

Muscat

Bonny Doon (Santa Cruz)	Robert Pecota (Napa)
Louis Martini (Napa)	Andrew Quady (Central)

These producers make a high quality, sweetish wine full of perfumed fragrance and the flavour of fruit.

Black grapes

Cabernet Sauvignon

Opus One (Napa)	Almadén (Santa Clara)
Lyeth (Sonoma)	Diamond Creek (Napa)

Château Montelena (Napa)
Beaulieu Private Reserve Georges
    Latour (Napa)
Ravenswood (Sonoma)
Clos du Bois (Sonoma)
Ridge (Santa Clara)

Heitz Martha's Vineyard (Napa)
Spottswoode (Napa)
Inglenook Reserve (Napa)
The Christian Brothers (Napa)
Rubicon (Napa)
William Hill Reserve (Napa)

These are beautiful, opulent wines of exceptional balance and generous flavour which would grace any table or banquet.

Merlot

Clos du Val (Napa)
Inglenook (Napa)

Jordan Winery (blended with
    Cabernet Sauvignon) (Sonoma)

These wines are high in alcohol and big and heavy in style. They are often blended with *Cabernet Sauvignon* to achieve a softer, rounder finish.

Pinot Noir

Robert Mondavi Winery (Napa)
Kalin (Marin)
Saintsbury (Napa)

Acacia (Napa)
Hanzell (Sonoma)
Clos du Bois (Sonoma)

An improving style of wine, at its best when rich and rounded, with a smooth, toasted vanilla, oak flavour.

Syrah

Bonny Doon (Santa Cruz)
Qupé (Santa Barbara)

Santa Cruz (Santa Cruz)
Joseph Phelps (Napa)

A light, fruity wine in the style of Beaujolais.

Zinfandel

Ridge (Santa Clara)
Fetzer (Mendocino)
Storybook (Napa)

Ravenswood (Sonoma)
Stevenot (Calaveras)
Whaler Vineyards (Mendocino)

Zinfandel is a multi-purpose grape making blush and pink wines which are then called white. This is California's native grape and its main reputation lies with the bold bright reds, rich in harmony and balance.

— New York State —

The state of New York produces 10% of the wine consumed in America. The extremes of climate make wine production precarious, and vines of the genus *Vitis labrusca* (such as *concord*, *catawba*, *delaware*, *dutchess*, *ives* and *niagara*) seem to thrive and do better than the great *Vitis vinifera* family. The wines have a very forthright, forceful flavour, not to everybody's taste. Sometimes, to make the wine more generally acceptable, it is blended with the softer, smoother Californian wines. Things are changing, and white wines made from the *Vinifera* grapes (*Chardonnay*, *Sauvignon*, *Riesling* and *Gewürztraminer*) are emerging as good styles in their own right.

The best producers of New York State wines are Vinifera Cellars, Bridgehampton, Hargrave, Pindar, Lenz, Wagner and Weimer. The major areas are **Finger Lakes**, **Long Island** and the **Hudson River**.

— Oregon —

Large quantities of wine are made here in the districts of the **Wilamette Valley** and **Roseburg**. The great grapes here are the *Chardonnay* and the *Pinot Noir*, although the *Riesling*, *Pinot Gris* and *Muscat* vines are also cultivated. There are great visions and hopes for the temperamental *Pinot Noir* and the prediction is that this grape will quite soon make really classy red wines here.

Best producers in Oregon are Amity, Peter Adams, Elk Cove, Rex Hill, Sokol Blosser, Adelsheim and Bethel Heights. From the *Pinot Noir Blanc* the Shafer Winery makes a really good blush wine.

— Best American sparkling wines —

The best American sparkling wines are:

Paul Masson (Santa Clara)	Great Western (New York State)
Schramsberg (Napa)	Domaine Chandon (Napa)
Gold Seal (New York State)	Iron Horse (Sonoma)

These are all made by the *méthode Champenoise* and some are even marketed in the United States as Champagne.

YUGOSLAVIA

Since the 1950s Yugoslavian white wines have been prominent and popular in Britain – who has not been weaned on Yugoslav Riesling, Lutomer Riesling, Laski Riesling and the sweeter Tiger Milk? More recently, newer wines like Lutomer Gewürztraminer and Lutomer Sauvignon, Beli Burgundec, Zilavka and Fruska Gora are establishing reputations.

Red wines are also putting on a show, especially the Merlot and Cabernet Sauvignon styles. Also look for names such as Prokupac, Dingač, GRK, Blatina, and Crnogorski Vranac ('Vranac' means 'black stallion'). Opolo is the best rosé wine; others are Cvicek and Kavadarka.

4 Other alcoholic drinks

AROMATIZED WINES AND BITTERS

— Vermouth —

The name *vermouth* is derived from the German *Wermutwein* – a wine flavoured with wormwood which, because of its therapeutic and digestive properties, was once highly esteemed as a medicine.

Historically the production of vermouth was based in **Turin** in Italy and **Marseille** in France. Traditionally Italy produces the sweet red pungent style and France the light dry white variety. Today there is no such style demarcation as both these and other countries make a variety of vermouths.

Vermouth is made from about 50 different ingredients. These include fruit, roots, barks, peels, flowers, quinine and a variety of herbs. The herbs are specially chosen for their aroma and medicinal properties. The basic wine used is ordinary and placid rather than good. It is matured for about three years and then *mistelle* (unfermented grape juice with the addition of brandy) is added. Meanwhile, the flavouring agents are macerated or infused in alcohol. This flavoured spirit is then added to the wine mixture and the lot is thoroughly blended in large tanks. At this stage some tannin may be added to give more flavour depth. The liquid is then fined, filtered, pasturized and finally refrigerated to ensure that any remaining tartrates crystallize and fall to the bottom of the tank.

Vermouth must be rested for a short period before being bottled for sale. A short shelf-life is ideal as it does not improve in bottle and is meant to be drunk young.

Vermouth types

The four main types of vermouth are:

Dry vermouth: often called French vermouth or simply French; it is made from dry white wine that is flavoured and fortified.

Sweet vermouth/Bianco: made from dry white wine, flavoured, fortified and sweetened with sugar or mistelle.

Rosé vermouth: made in a similar way to Bianco but it is less sweet and is coloured with caramel.

Red vermouth: often called Italian vermouth, Italian or more often It (as in Gin and It); it is made from white wine and is flavoured, sweetened and coloured with a generous addition of caramel.

Popular brands

Cinzano Red	Martini Rosé		Martini	
Cinzano Bianco	Martini Rosso	sweet	Cinzano	dry
Martini Bianco	Noilly Prat Red		Chambéry	
			Noilly Prat	

The delicate, dryish *Chambéryzette* is made in the Savoy Alps of France and is flavoured with the juice of wild strawberries.

Punt-e-Mes from Carpano of Turin is heavily flavoured with quinine and has wild contrasts of bitterness and sweetness. It is the personality vermouth, people either love it or hate it. Try it on the rocks with a slice of orange. *Carpano* is a similar style of vermouth but less bitter to the taste.

Service of vermouth

Use a Paris goblet or any stemmed glass. Serve a 3 out measure either chilled or with ice, soda water, tonic water or lemonade. It mixes well with sundry spirits and is an important ingredient for many cocktails. A lemon slice is the garnish for the dry varieties and a cherry on a cocktail stick for the sweet styles.

— *Bitters* —

Bitters are used either as aperitifs or for flavouring mixed drink and cocktails. The most popular varieties are mentioned below.

Campari

Campari is most favoured as an aperitif. It has an alcohol strength of 25% by volume and is made from a blend of herbs, gentia, bitter orange peel and quinine all of which have been macerated in spirit. It is rich red pink in colour and is best served with ice, a slice of orange and topped up with soda water. Many prefer it with tonic water, lemonade, sparkling mineral water or orange juice.

Angostura bitters

Angostura bitters was once made in a town of the same name in Bolivia. The town is now known as Ciudad Bolívar and today Angostura is produced in Trinidad. It is made from gentian and vegetable spices and is known as the 'Worcester sauce' of the cocktail business.

Angostura is an essential ingredient in that famous British naval drink *'Pink Gin'* and for the classic *'Champagne Cocktail'*.

Byrrh (pronounced *beer*)

This is a style of bitters made in France near the Spanish border. It has a base of red wine and is flavoured with quinine and herbs and fortified with brandy.

Other well known bitters are *Fernet Branca*, *Amora Montenegro*, *Radis*, *Unicum*, *Underberg*, *Abbots*, *Peychaud*, *Boonekamp* and *Welling*. Many are used to cure that 'morning after the night before' feeling. Cassis or grenedine are sometimes added to make the drink more palatable. Orange and peach bitters are also essential ingredients for some cocktails.

— *Other aromatized wines* —

Dubonnet

Dubonnet is available in two varieties: blonde (white) and rouge (red) and is flavoured with quinine and herbs. It was first made by Joseph Dubonnet as a tonic but modern versions made in south west France have a wine base with mistelle, spirit and flavouring added.

St Raphaël

This red or white, bitter-sweet drink from France is flavoured with herbs and quinine. It is a style of *ratafia*, once the 'good will' drink offered when a legal document was signed or 'ratified'.

Lillet

Lillet is a very popular French aperitif made from white Bordeaux wine and flavoured with herbs, fruit peels and fortified with Armagnac brandy. It is aged in oak casks.

Pineau des Charentes

Although not strictly an aromatized or fortified wine, Pineau des Charentes has gained popularity as an alternative aperitif or digestif. It is available in white, rosé or red and is made with grape must from the Cognac region and fortified with young Cognac to about 17% alcohol by volume.

BEER

The term *beer*, Britain's national alcoholic beverage, covers all beer-like drinks such as ales, stouts and lagers.

— *Composition* —

Beer is made from a combination of *water*, *barley* or other grain, *hops*, *sugar* and *yeast*.

Water

Also known as *liquor*, this must be of the finest and purest quality and may be of the hard or soft variety.

Barley

Barley is the best grain for beer making. It gives unrivalled taste and is low in protein (which causes cloudiness).

Malted barley is required for brewing. The malting process has three distinct steps – *steeping*, *germination* and *kilning*. The object of the process is to produce starch and enzymes in the grain so that enzymic conversion of the starch to fermentable sugars can occur readily in the brewhouse. The barley is first screened, then steeped in water and later spread out on a malting floor which is air conditioned. In the germination process each grain begins to sprout and tiny rootlets appear; at the same time enzymes begin to convert the starch within the grain to sugars.

The main fermentable sugar obtained is *maltose*, with small amounts of *glucose*, *fructose* and *sucrose*. Non-fermentable dextrins such as *maltotriose* and more complex carbohydrates remain and go through into the finished beer.

When the malting has reached the desired stage, the barley is heated in kilns to arrest the process. Kilning reduces the moisture content to about 4.5%, allowing the malt to be stored safely. It is important that the kilning temperatures used do not destroy the all-important enzymes which developed during germination. The malted barley is now ready for brewing.

Hops

These come from the same family as nettles but more specifically from the species *Humulus lupulus*. The two main styles for brewing are *Golding* and *Fuggles*. Hops give flavour, aroma and preservatives to beer.

Sugar

The sugar used in brewing is of the *invert* variety as it facilitates fermentation. In the form of a priming solution it is used to sweeten some beers.

Yeast

Yeast is needed for fermentation. It is specially cultured: *Saccharomyces carlsbergensis* for lager and *Saccharomyces cerevisiae* for other beers.

Finings

Traditionally *isinglass* (the bladder of the sturgeon) is used as the fining agent to make beer brilliantly clear.

— *How beer is made* —

Malted barley and a mixture of flaked and roasted barley (the degree of roasting dictates the final colour of the beer) is mixed with hot water and the combination is known as the *wort*. This is now taken to huge copper containers where hops are added. The hopped wort is cooled and transferred to a fermenting vessel where yeast is added. Fermentation takes place and after about seven days the liquid emerges as beer. It now stands in the storage vats and may be fined or pasteurized before leaving the brewery in a variety of containers.

Reduced alcohol beer

There are two categories of beer with reduced alcohol levels:

(a) *non-alcoholic beers* (NABs) which, by definition, must contain less than 0.5% alcohol by volume
(b) *low alcohol beers* (LABs) which, by definition, must contain less than 1.2% alcohol by volume

Beer is firstly made in the traditional way and then the alcohol is removed. The two favoured processes for removing the alcohol are *vacuum distillation* and *reverse osmosis*.

Vacuum distillation In vacuum distillation the alcohol and other volatile compounds are removed by passing the beer down a heated column under conditions of vacuum. These lower the temperature needed to evaporate off the alcohol: the lower the temperature the better will be the flavour of the ensuing product.

Reverse osmosis Reverse osmosis, also known as *cold filtration*, is carried out at low temperature. The beer is passed through a system of cellulose membranes which permit small molecules of alcohol to pass through.

Both of these techniques can be used to produce non- and low alcohol products. It is also possible to produce low alcoholic products by arresting fermentation at the appropriate point. This however tends to give a strongly flavoured product which is not particularly close to the flavour of a finished beer.

— Beer types —

Bitter: pale, amber-coloured beer served on draught. It may be sold as light bitter, ordinary bitter or best bitter (the latter usually being the brewer's premium brand).

Mild: can be light or dark depending on the colour of the malt used in brewing. It is sold on draught and has a sweeter and more complex flavour than bitter.

Burton: strong, dark, draught beer especially popular in winter when it is often mulled or spiced and offered as a winter warmer.

Old ales: these are brown, sweet and on the strong side. Again, they are ideal for mulling and spicing or simply drunk on their own.

Strong ales: these vary in colour between pale and brown, in alcohol between strong and very strong, and in taste between dry and sweet.

Barley wine: traditionally an all-malt ale, sweet, strong and swift to have effect. It is sold in stubby bottles or nips (equivalent to ⅓ pint). The alcoholic strength is about the same as a double whisky.

Stout: made from scorched, very dark malt and generously flavoured with hops. It has a smooth, malty flavour and creamy consistency. It is sold on draught or in bottles, with traditionalists favouring it being served at room temperature.

Porter: this ale gets its name from its popularity with porters working in Dublin and London. It is brewed from charred malt, could be described as a cross between stout and bitter, and is highly aromatic and flavoursome.

Lager: this beer gets its name from the German *lagern*, meaning 'to store' and was originally made in central Europe. During production, the yeast (*Saccharomyces carlsbergensis*) ferments at the bottom of the fermenting vessel. The beer is stored at low temperatures for up to six months after fermentation in order to condition and mature it before it is sold either in bottle or on draught. The long storage period also assists the beer to withstand variations of temperature without clouding.

Bottle-conditioned beers: also known as sediment beers because, by their very nature, they throw sediment in bottle where their conditioning takes place. After fermentation, the inactive yeast remains and forms sediment. Such beers require care in both storing and pouring. Prime examples of the style are Worthington 'White Shield' and Guinness Extra (in bottles not cans).

Other bottled beers: light ales, brown ales and export ales are pasteurized beers, free of sediment. They have an affinity with ordinary bitter, mild ale and best bitter respectively.

— *Beer strength* —

The average strength of beer is 4% alcohol by volume, but this figure varies with the type of beer produced. The tables below illustrate different types and makes of beer available, classified in terms of their strength.

SUPER STRENGTH (8–11% alcohol by vol)		
Lager	*Ale*	
Tennent's Super (9%) Carlsberg Special brew (9%)	Gold Label Strong Ale (10.9%)	

PREMIUM STRENGTH (4–6% alcohol by vol)		
Lager	*Ale*	*Stout*
Stella Artois (5.1%) Harp Premier (5%) Foster's (5%) Beck's Bier (5%) Harp Extra (4.5%) Miller Lite (4.2%) Tennent's (4%) Carling Black Label (4%)	Ruddles County (5%) Stones Best Bitter (4.1%)	Guinness Extra (4.3%) Draught Guinness in cans (4.1%)

STANDARD STRENGTH (3–4% alcohol by vol)	
Lager	*Ale*
Castlemain XXXX (3.6%) Tennent's Pilsner (3.5%) Harp (3.5%) Heineken (3.4%)	Tetley Bitter (3.6%) Whitbread Best (3.5%) Flowers Bitter (3.4%) Worthington's Special (3%)

LOW ALCOHOL (0.5–1.2% alcohol by vol)	
Lager	*Ale*
Tennent's LA (1%)	Whitbread White Label (1.2%)
Swan Light (0.9%)	Bass LA (1%)
Dansk LA (0.9%)	
McEwan's LA (0.9%)	
ALCOHOL-FREE (not more than 0.5% alcohol by vol)	
Lager	*Ale*
Kaliber (0.5%)	Smithwick's AFB (0.5%)

— *Service of beers* —

To allow for a good, attractive head or collar, glasses should be larger than the amount of beer to be dispensed. Under the EEC hygiene regulations a clean glass should be used for each new order. This does not please people who prefer their old glasses to be refilled. Glasses should be clean and brilliant in appearance, free from chips, cracks, smudges such as lipstick marks and other blemishes. They should be handled by the base, around the centre or by the handle – never by the rim.

Draught beers are dispensed in four ways:

(a) by a manual pull-beer engine
(b) through a free-flow tap
(c) through measured beer dispensers
(d) through taps with a back-action creaming device.

Serving temperatures

Beers are usually served within the temperature range 12.5–15.5 °C (55–60 °F) with lagers generally cooler than other beers. However, some people prefer to drink certain stouts at room temperature (18 °C/65 °F).

Pouring

Hold the glass at an angle and control the head of the beer by pouring against the inside of the glass. Lower and straighten the glass when a head needs to be encouraged. Never allow the tap or bottle neck to come in contact with the beer when pouring.

Certain bottle beers such as light ales and pasteurized beers can be poured straight into the glass. Others which have been *bottle-conditioned* must be poured very carefully. A typical example of a *sediment beer* is Worthington 'White Shield'. The skill is not to disturb the sediment when pouring so that only the brilliance of the beer can be seen in the glass: hold the tilted glass at eye-level and pour the beer very carefully down the inside of the glass, keeping the bottle absolutely steady. As the glass fills, lower it from the bottle to ensure an attractive head. When the sediment reaches the shoulder of the bottle, stop pouring.

The perfect glass of beer should look good enough to photograph, should be 'star-bright' in appearance, should taste true to type and should lace the glass (with froth) as it is drunk.

Faults in beer

Although thunder has been known to cause a secondary fermentation in beer thereby affecting its clarity, faults can usually be attributed to poor cellar management.

Cloudy beer This may be due to too low a temperature in the cellar or, more often, may result from the beer pipes not having been cleaned properly.

Flat beer Flat beer may result when a wrong spile has been used – a hard spile builds up pressure, a soft spile releases pressure. When the cellar temperature is too low, beer often becomes dull and lifeless. Dirty glasses and those that have been refilled for a customer who has been eating food will also cause beer to go flat.

Sour beer This may be due to a lack of business resulting in the beer being left on ullage for too long. Sourness may also be caused by the filthy habit of adding stale beer to a new cask or by beer coming in contact with old deposits of yeast which have become lodged in the pipeline from the cellar.

Foreign bodies Foreign bodies or extraneous matter may be the result of productional or operational slip-ups.

— *Beer outlets* —

Once the beer leaves the brewery it is sold in two types of outlets: *free houses* and *tied houses*.

Free house

This is a licensed premises which is privately owned and has no attachment to one particular source.

Tied house

This is a licensed premises which is either *tenanted* or *managed*.

Tenanted The tenant leases the property from the brewery and is tied to that brewery for the purchase of beer and perhaps other drinks. The tenancy agreement lays down the conditions of operation.

Managed A manager is paid a salary to run the premises for the brewery.

CIDER AND PERRY

Cider is fermented apple juice and is synonymous with counties such as **Somerset**, **Dorset**, **Devonshire** and **Herefordshire**. The apples are crushed and pressed, and cultivated yeast is added to the juice which ferments, resulting in cider. Dry ciders are completely fermented, but fermentation is stopped at a certain stage for the sweeter varieties. The cider is then matured for a few weeks before it is sold. Excellent cider is also made in France, especially in Calvados country (**Normandy**).

Pomagne

Pomagne is a sparkling cider made through the natural production of CO_2 by a secondary fermentation in the bottle. It was once known as Champagne cider but can no longer be so promoted.

Scrumpy

This is a name for strong, homemade, rough cider.

Perry

Perry is the name given to fermented pear juice made after the style of cider.

— *Popular brands* —

The following are the most popular brands of cider:

Bulmers   Coates
Gaymers   Whiteways
Merrydown

SPIRITS

Spirits are also known as *uisge beatha*, *eau de vie* or *aqua vitae*, all meaning *water of life*. They are all distillations of fermented liquids and are based on the principle that alcohol boils at 78 °C and therefore vaporizes before water (which boils at 100 °C). So, when a fermented liquid is heated to over 78 °C, the alcohol vapours rise and are drawn off into a condenser where they are converted into liquid raw spirit. It takes ten bottles of wine to produce one bottle of brandy so it can be seen that distillation is really a process of *strengthening by reduction*.

The two main methods of distillation are the *pot still* and the *patent still*.

— *Pot still* —

The pot still produces all the heavier spirits like brandy, malt whisky, calvados and dark rums. This method retains most of the *congenerics* during distillation. These flavouring agents are essential constituents to the flavour and personality of the spirit.

— *Patent still* —

The patent still, also known as the *Coffey still*, was invented in Dublin in 1830. It makes the lighter spirits such as gin, vodka, light rums, and light whiskies for blending.

— *Proof* —

The term *proof* originated from the days when the strength of spirits was tested by mixing them with gunpowder. When ignited if the mixture barely flamed it was *underproof*, if it exploded it was *overproof*, but if it had a mild blue flame it was *proved* safe for drinking, hence the word. 100 ° is proof, anything under is underproof (UP) and anything over is overproof (OP).

There are three scales of measurement for gauging the amount of alcohol in a liquid.

1 OIML

The European method of measuring alcoholic strength was invented by a French physicist, Gay-Lussac (1778–1850). He expressed 0 °C as the absence of alcohol and 100 °C as pure alcohol. On the scale alcohol strengths are stated as the percentage of alcohol by volume in a drink at 15 °C. The modern European scale based on this principle is now known as *OIML* (*Organisation Internationale Météorolgique Légale*) and is becoming mandatory in the EEC.

2 Sikes (Sykes)

The British and Commonwealth method was invented by Bartholomew Sikes in the early part of the nineteenth century. He used an hydrometer to calculate the density and strength of an alcoholic liquid. He determined pure alcohol as 175 °. Most bottles of spirits sold in Britain show 70 ° on the label so the calculation:

$$\frac{70° \times 100}{175°}$$

will give you the percentage by volume, in this case 40%.

3 American system

The American method has pure alcohol at 200 ° – twice that of the European (*OIML*) scale. Thus, if you see 90 ° on a label, you must simply divide by two to get the percentage by volume (ie 45%).

From the above, it becomes clear that:
70 ° (Sikes) = 40 ° (*OIML*) = 80 ° (American)
The quicker we all adopt the *OIML* scale, the better.

The following table should simplify matters further:

British (Sykes)	American	European (Gay-Lussac) (OIML)	Alcohol % by vol
175	200	100	100
100 (proof)	114	57	57
70	80	40	40

— Bases for spirits —

The bases used in the most common spirits are listed in the table below. In each case the base is a fermented liquid.

Spirit	Base
Whisky, gin and vodka	Barley, maize or rye (ie beer)
Brandy	Wine
Calvados	Cider
Rum	Molasses
Tequila	Pulque

— Brandy —

Although there are many fruit brandies on the market, strictly speaking brandy is the distillation of *wine*. The name comes from the Dutch *brandewijn* (burnt wine). *Burning* was the ancient term used for distilling. Brandy is made in many countries, notably **France**, **Italy**, **Spain**, **Germany**, **Portugal**, **Australia**, **USA**, **South Africa**, **Greece** and **Cyprus**. Of all the brandies, *Cognac*, made in the **Charente** department of France, is best.

Cognac

Cognac is a protected name and the spirit is made from grapes grown in six distinct regions. In descending order of quality these are:

Grande Champagne   Fins Bois
Petite Champagne   Bons Bois
Borderies   Bois Ordinaires and Bois Communs

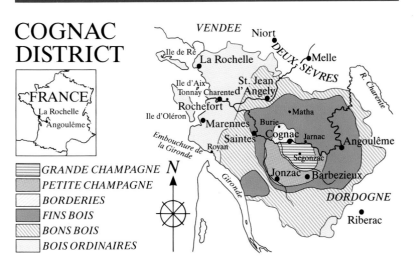

COGNAC DISTRICT

FRANCE
La Rochelle
Angoulême

- GRANDE CHAMPAGNE
- PETITE CHAMPAGNE
- BORDERIES
- FINS BOIS
- BONS BOIS
- BOIS ORDINAIRES

Although eight species of grapes are allowed by law to be used in the making of Cognac, only three (all white) are in fact used: *Saint-Emilion*, *Folle Blanche* and *Colombard*. They make very ordinary, harsh wine which only reveals true greatness when distilled.

The distillation takes place in copper *pot stills*. When the temperature passes 78 °C, alcoholic vapours rise into a pipe called a *swan's neck* (because of its shape). This carries the vapours into a condenser where they are changed into a liquid raw spirit. This first distillation is called *première chauffe* and is made up of:

(a) the head (*tête*) – the first part of the liquid to emerge. It is very pungent and unpleasant
(b) the heart (*brouillis*) – the middle and best portion of the distillate. This is taken to one side to be re-distilled – it has an alcoholic strength of 30% by volume
(c) the tail (*queue*) – the final liquid to emerge. Besides being very low in alcohol (water vapours have mingled at this stage), it is full of impurities.

The *tête* and *queue* are taken aside and later added to any new wine about to be distilled in order to purify them by distillation.

The *brouillis* is now distilled and this second distillation, known as the *bonne chauffe*, is called locally *la vigne en fleur* – the vine in flower. It has an alcoholic strength of 70% by volume and is full of fiery flavour as well as numerous trace elements.

Maturation The young spirit (*bonne chauffe*) is put into special wooden casks made of Limousin oak – the wood adds its own character and colour to Cognac. As maturing continues, the spirit is checked from time to time. If the quality is exceptional it is awarded *RN* (*Reserve Nouvelle*) status and put into a section of the cellar known as the *paradise* where all the vintage Cognacs mature to a great age. These are used later to help the quality and style of less fine products. (It is very unusual to see a bottle of vintage Cognac.)

Quality is helped by ageing in cask – the longer the better – because Cognac, like all spirits, improves only while maturing in cask, it does not improve with age in the bottle. An annual evaporation loss of 3% – known as the 'angels' share' is common and so casks are topped up each year with younger brandy to allow for this.

Brandies of great age and refinement are, of necessity, very expensive and are known as *liqueur brandies*. They should not be confused with *brandy liqueurs* which are liqueurs with a brandy base. Age is indicated by stars and words on the label, although these may be interpreted differently by different producers.

Label language		
	*	3 years maturing in cask
	**	4 years in cask
	***	5 years in cask
	VO	(very old) 10–12 years in cask
	VSO	(very superior or special old) 12–17 years in cask
	VSOP	(very superior or special old pale) 20–25 years in cask
	VVSOP	(very very superior or special old pale) 40 years in cask
	XO	up to 45 years in cask
	Extra	70 years in cask
	Fine Maison	brandy of the house
	Fine Champagne	Cognac made from Grande and Petite Champagne grown grapes
	Grande Fine Champagne	Cognac made only from Grande Champagne grapes

Popular brands The following are the brand names most frequently associated with Cognac:

Hennessy	Rémy Martin	Bisquit
Martell	Prince Hubert de Polignac	Courvoisier
Camus	Hine	Otard

Armagnac

From the **Gers** department to the south-east of Bordeaux comes the world's second best brandy – *Armagnac*. The grapes *Picpoule* (*Folle Blanche*), *Saint-Emilion*, *Jurançon* and *Colombard* make a basic white wine which is produced in three areas: **Bas-Armagnac** (the best area), **Ténarèze** and **Haut Armagnac**. This wine is distilled in a special type of *continuous still* and aged in black oak barrels for up to 20 years.

For real quality look for *Grand-Bas-Armagnac* or *Grand Bas* on the label. The best brands are Janneau, Sempé, Marquis de Montesquieu and Marquis de Puységur.

Other well-known brandies

Brandy is made in a variety of different countries – in fact, it can be produced anywhere that wine is made. Below is a list of the most well-known brandies available, other than Cognac and Armagnac. They are usually lower in quality and less expensive than these two brandies.

Asback Uralt (Germany)
Christian Brothers ⎤
Korbel ⎬ (USA)
Paul Masson ⎦
Fundador ⎤
Lepanto ⎥
Conde de Osborne ⎬ (Spain)
Bobadilla 103 ⎦
Pisco (Chile)

Vecchia Romagna ⎤ (Italy)
Stock ⎦
Oude Meester (South Africa)
Cambas ⎤ (Greece)
Metaxa ⎦
Anglias ⎤
Five Kings ⎬ (Cyprus)
Peristiani VO 31 ⎦

The following are brandies made from grape residue pressings which, when fermented and distilled, produce a fiery brandy full of personality that sends search lights through your body – but oh! the after glow!!

Aguardiente (Spain)	Bagaciera (Portugal)
Marc (France)	Grappa (Italy)

Service of brandy

Brandy should be served at room temperature in thin glasses that can be easily warmed in the hand to enhance the bouquet and flavour. There is no need for the gimmicky glass heaters which you see sometimes in 'with it' places. Brandy balloons are the best glasses to use as they curve in narrowly at the top to contain the aroma. Over-large balloons are wasteful, pretentious and make the measure appear stingy.

The common measure is ⅙ of a gill or multiples. Ginger ale, soda water or 7-Up may be added to brandy, but only the cheaper varieties of spirit should be so adulterated.

— *Calvados* —

Calvados is an apple brandy, the best quality of which is produced in the **Pays d'Auge** in Normandy. Basically it is distilled cider made in *pot stills* and aged in wood, sometimes for up to 20 years. Good quality Calvados like Un Trou Normand is a delightful drink.

In France it is customary to serve Calvados instead of, say, a sorbet mid-way through the meal to cleanse the palate and aid digestion. More usually though, it is one of the many digestifs offered in a restaurant at the end of a meal. The American equivalent to Calvados is *applejack brandy*, of which the brand Laird's is best known.

Service is as for all styles of brandy – in a small brandy balloon which can be heated in the hand easily.

— *Gin* —

Gin was first made in 1577 by a Dutch professor of chemistry called Sylvius Van Leyden. It was then called *genever* after its main flavouring agent the juniper berry. It was purposely made as a medicine because juniper berries have oils and other properties which were thought to be curatives against such illnesses as gout and kidney trouble. It was sold in chemist shops and doctors began to prescribe it, as some do to this day, for ailments such as prostate trouble.

English soldiers fighting battles in the Low Countries also liked it, especially before battle, and named it 'Dutch Courage'. Returning to England the soldiers made the drink popular under the name of *geneva*. The English soon began to make it themselves and shortened the name to *gin*.

In 1688 Dutch William of Orange came to the English throne and from then onwards gin became known as the Englishman's spirit. Gin was so lightly taxed in those early days that even the poorest could afford it. A drink could be obtained for a penny, and people could get dead drunk for tuppence – something that often occurred as the poor attempted to forget, for a while, their appalling social conditions. Unlicensed dram shops sprung up everywhere to the extent that, by 1729, one in every four houses in London distilled gin (of the foulest quality imaginable). Gin soon got the nickname 'mother's ruin' and 'royal poverty', and was sold in shops, street corners or wherever there was a hole in a wall. The rise in mental illness was enormous, as was the death toll. Several Acts of Parliament were passed increasing taxes in the hope that higher prices would have a sobering effect on the masses. Eventually more responsible people entered the business, improving the quality of the product, and gin took on a respectable image.

In the 1880s the Americans, with their gin-based drinks and forty years later with their cocktails, ensured the acceptability of the product. Today gin is a big seller worldwide. Who can resist a gin and tonic or a hefty dry martini cocktail after a hard day at the office?

How gin is made

Gin is a *rectified* (redistilled) spirit made mostly from maize, malted barley and rye. It is flavoured with a mash of juniper berries, cassia bark, coriander seeds, angelica root, calamus root, cardamoms, almonds, fennel and bitter orange peel. Initially an almost pure spirit is purchased from a central distillery. This is reduced in strength by the addition of distilled water and rectified (re-distilled) in a *patent still*, together with the flavouring ingredients. The resulting vapours are condensed and run off as a high strength gin. The heads and tails are returned for distilling with the new batch.

The other method of imparting flavour to the raw spirit is by steeping the *botanicals* (flavourings) in a proportion of raw spirit and then distilling it in a pot still. The result is then added to the rectified spirit produced by the patent still. It is rested for one month in oak casks and then tested for taste and clarity and reduced to the selling strength of 70 ° proof (40% alcohol by volume).

Gin styles

London dry gin This was originally made in London, hence the name. It is now made elsewhere under licence. Gordons, Booths, Beefeaters, Burroughs, Gilbeys, Seagrams and Tanqueray are typical examples.

Plymouth gin Made originally by Coates & Co in Plymouth. It is more aromatic than London Gin and is the standard gin used in the making of Pink Gin.

Dutch/Holland gin Also called *geneva* gin, it is made in a pot still which retains most of the flavours. It has a very definite personality and malty flavour which is why, in Holland, it is often followed by a lager chaser. This type of gin is made principally in Amsterdam and Schiedam. Bols, De Kuyper and Jonga Graan Jenever are all well-known brands.

Old Tom This is the name given to gin sweetened with sugar syrup. It is now made in Scotland and mostly for the export market. Traditionally Old Tom was the gin used in a Tom Collins cocktail. It gets its name from the time of the Gin Act of 1736 when a government informer purchased a picture of a cat and nailed it to the ground floor window of a rented house. Underneath the cat's paw was a lead pipe which was connected to a gin supply inside. Locals soon came to recognize the sign which was changed from location to location to avoid discovery. Twopence in the cat's mouth secured a good mouthful of gin.

Sloe gin This is made by steeping sloes (the fruit of the blackthorn) in London dry gin, together with sugar syrup and a small amount of bitter almonds.

Fruit gins These are mostly artificially flavoured with the appropriate essences, eg orange or lemon.

Service of gin

Gin should be served in a Paris goblet large enough to hold the accompanying mixers, ice and garnish. The gin should be poured into the glass first, to ensure the right measure.

Some people drink gin neat or with a drop of Angostura bitters whirled around the glass (*Pink Gin*). Some like it with water or with dry vermouth (*Martini*), but the great majority prefer it with tonic water or bitter lemon, with the addition of ice and a slice of lemon.

— *Rum* —

Rum is distilled from molasses or sugar cane juice. Although the **Caribbean** is its true home it is made wherever sugar cane grows in abundance. In its early days it was popularly known as *rumbustion* or *rumbullion* (kill devil) because it was rough, crude and unmatured.

How rum is made

Fresh cut sugar cane is crushed in roller mills. The extracted juice is then boiled and, as the water evaporates, concentrated sugar is left behind. This is then clarified to become a thick, heavy syrup which is pumped into centrifugal machines. These crystallize the sugar and separate it from the molasses. Water is then added to the molasses which undergoes either a slow or quick fermentation. (The fermented liquid has an alcoholic strength of 7% by volume.)

Slow fermentation This can last for up to 20 days and produces all the big, heavy rums, eg Jamaican rum. The slow fermented liquid goes into a *pot still* where it is double distilled to produce a rum that is rich in *congeners* (esthers, acids, flavour and bouquet agents). It will have an alcoholic strength of 150° proof (86% by volume).

Quick fermentation This lasts for two to three days and is associated with the light rums such as Puerto Rico and Cuban rums. The quick fermented liquid is placed in a *patent still* for distillation. The resulting rum has very few *congeners*, is much lighter in body than that produced from the slow fermented liquid, but has a higher alcoholic strength (160° proof, 91% by volume).

Most rums sold in Britain are matured for at least three years in cask. Often they are transported away from the Caribbean (because the intense heat evaporates the liquid too quickly) to spend their maturing period in cellars belonging to the United Rum Merchants at Dundee. Before bottling the rum is reduced in strength by the addition of distilled water to 70° proof (40% alcohol by volume). Dark rum may have caramel added to ensure colour consistency.

Basically there are three styles of rum:

Dark rum The dark, sweetish, heavy-bodied rums come from Jamaica and Martinique.

Light rum This is pale in colour, drier and lighter in body than the dark rums, and comes from Cuba, Puerto Rico and Barbados.

White rum This rum comes from Puerto Rico, Cuba and Martinique and forms the base for such cocktails as Daiquiri and Cuba Libre.

Rum, of course, is also made in many other countries like Bolivia, Australia, South Haiti, South Africa and the USA.

Service of rum

Rum should be served in a tumbler, Paris goblet or highball glass. It is usually drunk neat, with ice or with mixers such as cola or 7-Up. Garnishes for rum, rum-based cocktails and punches include lime, lemon, orange, pineapple and cherries. The choice is dependent on the simplicity or complexity of the presentation.

Popular brands

Dark rums	White rums
Barbancourt	Barilla
Myers	Bacardi
Captain Morgan	Ron Rico
Lambs Navy	Rhum St James
Appleton	Dry Cane
Mount Gay	
Woods	
Hansen	

— *Tequila* —

Tequila is made from the sour juice of the *agave* cactus plant, which cannot be harvested before it is 12 years old. After fermentation the *pulque* is twice distilled in pot stills. After the first distillation is is known as *mescal* and can be sold at this stage if required. However, the product is very rough and fiery.

The second distillation refines the spirit to what we know as *tequila*. It is then aged in cask for some time before being bottled for sale. It is either white or golden in colour and is drunk accompanied by salt on the back of the hand and a wedge of lime. Lick the salt, drink the drink and suck the lime. Best brands are José Cuervo and Montezuma.

— *Vodka* —

While some vodkas are made from potatoes and others from grain, it can be made from any fermentable liquid. Vodka originally came from **Poland** and **Russia** where it was presented under such names as *wodka, wodki, votku, votky* and *vodka*, all meaning 'little water'. In fact, it looks like water, being colourless, tasteless and odourless – the latter quality has gained it a reputation as a 'new wife's deceiver'.

How vodka is made

Vodka is distilled and rectified in a *patent still*. The resulting almost pure spirit is slowly passed through layers of vegetable charcoal and activated carbon. This process removes all traces of colour and flavour. Because it is now a neutral spirit, it does not require maturing and when it is reduced in strength to 65.5 ° proof (37.4% alcohol by volume) for the British market, it is bottled and ready for sale. Other countries require a higher alcohol content.

Having said that vodka is flavourless, there are other styles of vodka which have flavour added through herbal infusion. For example, the famous Polish Vodka Zubrówka is quite aromatic because *zubrówka grass* (the grass the European bison feed on) is used in the infusion. Each bottle will have a blade of grass floating inside. It has a slightly nutty flavour and a green tinge.

Popular brands

The most popular brands of vodka available are:

Wyborowa Zubrówka	} Poland	Muskovskaya Stolichnaya	} Russia
Saratof Nordoff	} Ireland	Cossak Smirnoff	} England
Steinhäger Schinkenhäger	} Germany	Absolut	Sweden

Service of vodka

Vodka is best served neat in small tumblers which have come straight from the freezer ('with a tear', as the Russians say). The outside of the glass has therefore become frosted over with the intense cold. Served like that, vodka is best drunk in one gulp (no need to smash the glass).

A more leisurely way to drink vodka is with ice and a mixer in a Paris goblet or with fresh orange juice (*Screwdriver*), tomato juice (*Bloody Mary*) or with ginger beer (*Moscow Mule*).

— *Whisky* —

The name 'whisky' comes from the Gaelic *uisge beatha*. Christian monks brought the art of distilling from Ireland to Scotland and, although the oldest registered distillery is Bushmills (1608) in County Antrim, the real credit for elevating whisky to the position of being the most internationally enjoyed spirit must go to the Scots. It was their vision which saw the possibilities of blending grain and malt whiskies, the results of which are now so successful on world markets. Today many of the high flying commercial whiskies are a blend of 50% malt and 50% grain whiskies.

Malt whisky

Malt whisky is made by grinding *malted barley* (see page 219). To this is added hot water and, from this mash, a sweet liquid known as *wort* is extracted. Yeast is added to the wort and fermentation produces a *wash* with 7–10% alcohol (beer without the hop flavour). The wash now goes into a *pot still* and the resulting *low wines* are redistilled to produce malt whisky, high in alcohol.

The raw colourless whisky is put in oak casks (old rum and sherry casks are sometimes used as they add flavour and colour to the final product). The whisky matures in these casks for a minimum of three years although it may remain for up to 50 years. The length of maturation must be revealed accurately on the label. Once bottled it will not improve further.

Styles of malt

Single cask malt This is a specially-selected single malt which is matured in a particular cask for a specified number of years. It is then usually presented in a commemorative bottle to mark a special event or occasion. These malts are known in the trade as 'specials'.

Single malt Sometimes known as straight malt, this is the product of one single distillery.

Vatted malt The marriage of single malts from different distilleries.

Grain whisky

Grain whisky is much lighter in body and flavour and is made using the *patent (Coffey) still* developed by Aeneas Coffey who was Inspector-General of Excise in Ireland. The patent still has two cylindrical columns: one an analyser, the other a rectifier. The wash is pumped into the top of the analyser where, as it descends, it meets a current of raw steam which strips it of its alcohol. The alcohol vapours pass through the rectifier while the used wash is taken out at the bottom. When the spirit vapours reach the top of the rectifier they are condensed on a water frame. The foreshots (*tête* – the first liquid to emerge) are taken and added to the new wash and the comparatively pure spirit which follows is drawn off into the receiver and held until required for blending.

Blending and ageing

Blended whiskies are made from a mixture of malt and grain whisky. The more malt in the blend, the better and more expensive the product. The length of maturation is also reflected in the price. The age on the label refers to the youngest whisky in the blend and indicates the length of time the whisky has matured in cask.

Scotch whisky

Of all the whiskies produced, *Scotch* is by far the biggest seller in the world today. The consistent quality associated with each brand results from five very important factors:

(a) The *shape* of the *pot still*, which retains maximum flavour and smell in the spirit.
(b) The *quality* of the *water*, which must be absolutely pure as must be the demineralized water used to reduce the strength to a potable level 70° proof (40% alcohol by volume).
(c) The *quality* of the *malt* which uses peat fires to impart the traditional reek or smoke flavour.

(d) The *care* and *attention* as the whisky matures in cask.

(e) The *skill* of the blender.

Popular brands

Popular brands of Scotch are:

Malt	Blended	De luxe
Auchentoshan	Bells	Dimple Haig
Cardhu	Teachers	Johnnie Walker Black
The Glenlivet	Black and White	Label
Glenfarclas	Cutty Sark	Chivas Regal
The Glenfiddich	Ballantine	The Antiquary
Glenmorangie	Sheep Dip	(These have a high
The Glenturret	Vat 69	percentage of malt
Knockando		whisky in the blend)
Laphroaig		
Highland Park		
Isle of Jura		
Talisker		

Blended Scotch accounts for about 90 per cent of all Scotch sold. The perfect blend is said to be a combination of Highland, Lowland, Island (Islay) malts and grain whiskies.

Irish whiskey

The Irish Distillers Group has, since 1966, taken over the marketing and promotion of Irish whiskey. The industry as a whole has benefited, especially with the emergence of Irish whiskey-based liqueurs and the popular whiskey-based post-prandial drink, Irish coffee. Irish whiskey uses *malted barley* and to a lesser extent *unmalted barley*, *wheat*, *rye* and *oats* in the *mash*. It is triple distilled and has more malt in the blend than some other blended whiskies. Quite recently Bushmills produced Ireland's first 'completely malt' whiskey which is supremely smooth and delicate. It is aged for at least five years in cask and very often up to 15 years.

Brand names

Coleraine	Bushmills 10 Year Old Malt	Black Bush
Jameson	Paddy	Middleton Very Rare
Tullamore Dew	Powers Gold Label	

American whiskey

The art of distilling was brought to America by Irish immigrants who settled in Pennsylvania in the middle of the seventeenth century. Their whiskey became so popular that the George Washington government decided to impose a tax on the product which led to the Whiskey Rebellion in 1794. The distillers refused to pay and went further into Western Pennsylvania, Kentucky, Georgia, Indiana, Illinois and Ohio and set about re-establishing their business. Whiskey became so popular that the Temperance Movement and others had it banned by the notorious Fourteenth Amendment of November 1920. Prohibition lasted for 13 long years until 1933 when the distillers were once again allowed to ply their trade.

Most of the whiskey produced is called *Bourbon* after Bourbon County, Kentucky, where it was first made. It is produced from a mash containing not less than 51% *maize* (*corn*) but more usually up to 70% maize. Famous examples are Jim Beam, Old Crow, Old Forrester, Old Grandad, Taylor, Wild Turkey and Four Roses. Jack Daniels, although not a Bourbon, vies with these in reputation and quality. It is made in Tennessee.

Other styles of American whiskey

Rye whiskey	made from a mash containing not less than 51% rye
Corn whiskey	made from a mash containing at least 80% maize. It is fiery and gets little ageing
Straight whiskey	unblended and made of one type of grain. It is aged at least two years in charred oak casks
Bottled in bond whiskey	straight whiskies which are matured in cask in government bonded warehouses for four years. They are bottled at 100 ° USA proof (50 % alcohol by volume)
Blended whiskey	a blend of straight whiskey and neutral spirit. Its light character makes it an ideal ingredient for some cocktails
Sour mash bourbon	is made from a mash which has yeast in it that was used in a previous fermentation. Most American whiskey is aged in charred oak casks which impart quality and colour.

> *The Irish and Americans spell the name of their product with an 'e': whiskey. All other countries leave out the 'e'.*

NB Southern Comfort is a whiskey-based liqueur flavoured with peaches and oranges from Georgia.

Canadian whisky

Canadian whisky is made mostly from rye and maize. Being distilled in a form of *patent still*, it is very light in character and is ideal for use in cocktails. It is aged in charred oak casks for a minimum of three years. Brand names include Seagram Canadian Club, Hiram Walker, Gilbey Canada, Wiser's Oldest and Canadian Schenley.

Service of whisky

Whisky is usually served in a Paris goblet or tumbler. Depending on the country ⅙, ⅕ or ¼ of a gill (¼ pt) is the measure used. Whisky is at its best when served neat, but personal preference dictates whether it is served on the rocks (with ice), with soda water, mineral water or tapwater, or with mixers such as lemonade, 7-Up, tonic water etc. Scotch whisky with ginger wine is *Whisky Mac*. Irish whiskey with ginger wine is *Whiskey Mick*.

— *Other spirits* —

Aquavit/Schnapps

Aquavit is the national drink of **Scandinavia**, but is also associated with **Germany**. It is nothing more than flavoured vodka, the most common flavouring being caraway, dill, cumin, bitter orange, fennel and aniseed. It is known as *schnapps* in Germany, or *snaps* in Norway and Sweden, and *schnaps* in Denmark. The name means gasp or snatch.

Aquavit should be drunk in one gulp in a small glass taken straight from the freezer. It is ideal with the cold food of Scandinavia and Germany. Best brands are Bommerlunder (Germany), Linie (Norway) and Aalborg (Denmark).

Korn

Korn, or kornbranntwein, is another variety of schnapps, and is a low strength brandy-style spirit, but made from grain.

Doppelkorn

Doppelkorn is a much stronger version of korn, often flavoured with juniper berries and sold under such well-known labels as Fürst Bismarck and Doornkaat. It makes a great aperitif and goes very well with hors-d'oeuvre fish dishes such as smoked salmon, sardines, roll mops and *matjesfilet*.

Arrack

Arrack is made from the fermented and distilled sap of the palm tree to which rice and molasses have often been added during fermentation. It is popular in **India, Sumatra, Bornio, New Guinea** and **Java** where the style Batavia Arrack is highly thought of.

Eau de vie

Eau de vie is the fermented and distilled juice of fruit, much of the best of which comes from the **Alsace** area of France, **Germany, Switzerland** and **Yugoslavia**. Examples are:

Himbeergeist	from wild raspberries (Germany)
Kirschwasser	from cherries (Germany)
Mirabelle	from plums (France)
Quetsch	from plums (Alsace and Germany)
Poire Williams	from pears (Switzerland and Alsace)
Slivovitz	from plums (Yugoslavia)
Fraise	from strawberries (France, especially Alsace)
Framboise	from raspberries (France, especially Alsace)

Eau de vie, especially the *alcool blanc* variety, should be water-clear in appearance and served well chilled.

Pastis

'Pastis' is the name given to spirits flavoured with *anis* and/or *liquorice*. The spirit is made in many Mediterranean countries and is popular almost everywhere. It has taken over from the infamous absinthe, once known as the 'Green Goddess'. The latter has since been banned in most countries since it has been found to cause serious illnesses, such as insanity and sterility. This results from its high alcoholic content plus the use of wormwood as the flavouring agent. Such a combination is lethal.

Pastis turns milky when water is added because the water brings the natural oils of the flavouring agents out of solution into suspension.

Popular brands are Pernod, Ricard, Ouzo, Raki, Ojen, and the non-alcoholic brands Blancard and Pacific.

Service Pastis should be served with iced water in a Paris goblet. The usual proportion is three parts water to one part pastis, but extra water can be added if it is preferred weaker.

Sake

Japanese *rice wine* is actually a rice-based spirit. It was first made in 712 AD but became commercialized in the eighteenth century. It is first brewed, then distilled, and is marketed in the traditional flask-like *tokkuri* bottles. It has an alcohol strength of about 17% by volume and is usually served in small ceramic cups called *sakazuki*. It may be drunk warm or cold.

LIQUEURS

Liqueurs are flavoured and sweetened spirits, sometimes artificially coloured to please the eye. The base spirit is normally *brandy, rum,* a *neutral spirit* or *whisky*. The flavouring agents can be *aniseed, caraway, almond kernels, nuts, soft fruits* (eg *cherries, blackcurrants, apricots, peaches, raspberries, strawberries, bananas*), *citrus fruit peel, gentian, wormwood, rose petals, violets, hissop, myrtle, peppermint, cinnamon, sage, coriander, nutmeg, angelica, honey, coffee beans, cocoa beans, coconut* and *cream.*

Many liqueurs originated in monasteries and were made initially either to disguise poor quality spirits or as a curative for colds and fevers. Even today they are made in many countries principally as an aid to digestion. Together with brandies and some other spirits, liqueurs come under the heading of *digestifs* on a wine list.

There are two methods of making liqueurs: the *cold* or *maceration method* and the *heat* or *infusion method.*

— *Maceration method* —

The maceration method is used when *soft fruits* or *resinous substances* provide the aroma and flavour. *Brandy* is usually the base spirit and this is put into oak casks with the fruit and kept at room temperature for up to a year, being stirred occasionally. Gradually flavour and colour are extracted from the fruit.

When maceration is completed, the liqueur is drawn off, filtered and, if necessary, adjusted for colour.

— *Infusion method* —

The infusion method is associated with flavour agents such as *herbs*, *roots* and *peels*. These are rich in oils which can be extracted, together with flavour and aroma, by heating.

The flavouring agents are soaked in the base spirit for a day. The mixture is then transferred to a *pot still* where, after distillation, the flavoured vapours are condensed as liqueurs.

Sometimes an apparatus working on the principle of a percolator is used. The flavouring ingredients are placed in the upper half and the spirit in the lower. As heat is applied, the alcohol vapour rise and seep through the ingredients, extracting all the flavours and essential oils. When condensed the liquid will be colourless. It may be left water-clear or artificially coloured for psychological and commercial reasons.

— *Popular liqueurs* —

Abricotine

Made near Paris in Enghien les Bains, this liqueur is made from apricots, their kernels and brandy.

Advocaat

Made mostly in the Netherlands by the firms Bols, Warninks and de Kuyper. Its ingredients include egg yolks, sugar, vanilla and brandy.

Anisette

Colourless, aniseed flavoured liqueur. The style produced by the French firm Marie Brizard is particularly good.

Atholl Brose

A kind of liqueur made in Scotland from malt whisky, honey, cream and fine oatmeal.

Bailey's Irish Cream

The biggest selling liqueur. Launched in 1975, it has Irish whiskey, chocolate, honey and Irish cream as its main ingredients. There are many other Irish creams on the market, but Bailey's is considered to be the best.

Bénédictine

The monastic order at Fécamp in Normandy first made this liqueur in 1510. It is a blend of brandy and some 75 herbs and spices including coriander, cardamom, cinnamon, hyssop, vanilla, saffron, mace and nutmeg. It was first made as a medicine to combat malaria but the monks have made it commercially for a long time. The initials *DOM* on the label stand for *Deo Optimo Maximo* (to God, most good, most great). *B & B* is a mixture of brandy and Bénédictine in equal quantities.

Brontë

Made in Yorkshire and named after the famous Brontë sisters. It has a base of French brandy and is flavoured with herbs, spices and oranges.

Chartreuse

Made by the Carthusian monks near Grenoble in France and Tarragona in Spain. It is brandy based and flavoured amongst other things with hyssop, cinnamon, angelica, saffron and mace. Chartreuse is available in two colours, green and yellow: the green has a dry, but powerful, flavour and is probably one of the strongest liqueurs available in alcohol terms (55% by volume); the yellow has 43% alcohol by volume and a honey-sweet flavour.

Cherry brandy

A very sweet and popular cherry-flavoured, brandy-based liqueur. Best brands are Cherry Heering, Cherry Rocher, Dolfi, Garnier and de Kuyper.

Coconut liqueurs

These have recently become very trendy. Malibu, with a white rum base, is a very big seller. Others are Cocoribe and Batida de Coco from Brazil.

Coffee liqueurs

These have always been popular, eg Tia Maria from Jamaica, Kahlúa from Mexico and Bahia from Brazil. They also all go exceptionally well when poured over ice cream. Tia Maria, which is rum based and flavoured with Blue Mountain coffee, is often served with cream floating on top.

Cointreau

Made in Angers in France by the Cointreau family. Originally named Triple Sec, it is water-white in colour and flavoured with oranges. It is especially good served 'on the rocks'.

Cordial Médoc

Brandy based and highly aromatic, being flavoured with old Claret. It is made in Bordeaux.

Crèmes

Crème de banane	banana flavour
Crème de cacao	chocolate flavour
Crème de cassis	blackcurrant flavour
Crème de menthe	mint flavour
Crème de noyau	almond flavour

Curaçao

Originally made in Amsterdam and flavoured with the dried peel of small bitter oranges from Curaçao in the Caribbean. The colour is either golden or a delicate blue. It used to be known as Triple Sec, a curious description because it tastes very sweet.

Cuaranta y Tres

Yellow gold in colour, this Spanish liqueur is made from 43 different ingredients including many herbs, banana and vanilla.

Cynar

This is one of Italy's favourite liqueurs. Made from the heart of artichokes, it can certainly help to reconcile the turbulence of a full stomach.

Danziger Goldwasser

Contains very tiny specks of gold leaf. It is flavoured with caraway, aniseed and herbs. Originally made in Danzig, Poland, it is very popular in Germany and Scandinavian countries. Danziger Silberwasser has silver flakes instead of the gold.

Drambuie

Scotland's best liqueur. It is made by the Mackinnon family from old malt whisky, heather honey and herbs.

Fior d'Alpi

A yellow coloured, Italian liqueur. It is sold in tall, narrow bottles that contain a small twig on which rock sugar crystals have formed (reminiscent of a Christmas tree).

Forbidden Fruit

A brandy based liqueur that comes from America and is flavoured with shaddock, grapefruit, orange and honey and is golden brown in colour.

Galliano

A golden coloured Italian liqueur. It is made from 30 herbs, and various berries, roots and flowers and sold in very tall, narrow necked bottles. Galliano is one of the main ingredients for the popular cocktail *Harvey Wallbanger*.

Glayva

A Scottish liqueur with a whisky base and flavoured with herbs, spices, oranges and heather honey.

Glen Mist

A Scottish liqueur with a whisky base, flavoured with herbs and spices.

Grand Marnier

Made partly in Cognac and Paris from the juice of Caribbean oranges and aged Cognac. It is excellent on its own but is also used to flavour sweet soufflés, *Crêpes Suzette* and duck *à l'orange*.

Irish Mist

Ireland's answer to Drambuie with, of course, Irish whiskey as the base.

Izarra

Made in the Basque country of France from Armagnac. It is flavoured with a multitude of herbs and is available in two styles: yellow at 43% alcohol by volume and green at 55% alcohol. The name means 'star'.

Jägermeister

A German herb liqueur, dark red in colour, that is presumably a favourite with the hunters after a hard day in the field.

Kirsch

A water-white liqueur from Alsace and south Germany. Is made from black cherries and their kernels, it is a good partner to fresh pineapple and fruit salad.

Kümmel

A water-clear liqueur native to East European countries. It has a vodka (neutral spirit) base and is flavoured with caraway and cumin seeds.

Kumquat

An orange flavoured liqueur made in Corfu, Greece, from the oval shaped fruit of the same name.

Mandarine Napoléon

A well-presented orange flavoured liqueur from Belgium. It is made from Spanish mandarins and Cognac.

Maraschino

A colourless liqueur from Italy that originated in Dalmatia, Yugoslavia. It is made from sour red cherries and almonds and is often used to flavour fruit salads.

Mead

May be termed a 'liqueur'. It is made from a combination of honey, malt, yeast and spices. It used to be called Metheglin in America.

Parfait Amour (Perfect Love)

A liqueur which is highly scented and purple coloured to infer passion. Flavoured with spices and citrus fruits.

Sambuca

An Italian liqueur flavoured with liquorice and elderberry. It is very popular especially with coffee at dinner. Traditionally 3 coffee beans should be put in a glass of Sambuca and the liqueur flamed to extract flavour from the beans. This also helps create a cosy intimate atmosphere.

Southern Comfort

A popular whiskey-based liqueur from America which is flavoured with peaches and oranges.

Strega (Witch)

An Italian liqueur from Benevento, Naples, which is made from 70 different herbs and citrus fruits. It is yellow in colour and orange flavoured.

Trappistine

A liqueur made by the Trappist Monks at the Abbey de la Grace de Dieu, France. It is made from Armagnac brandy and local herbs and is yellow/green in colour.

Van der Hum

A popular South African, yellow coloured liqueur. Made from a base of Cape brandy, it is flavoured with a kind of tangerine called naartje, nutmeg and herbs.

Vieille Cure

Originally made at the Abbaye de Cenon near Bordeaux, this liqueur was intended as a medicine but became popular as a digestif. It has a brandy base and is flavoured with 50 herbs.

— *Service of liqueurs* —

Ideally liqueurs should be served in small brandy balloons which are roomy enough to allow the bouquet of the liqueur to be properly appreciated. They can be served neat, 'on the rocks' or with crushed ice and a straw, as in the service of *crème de menthe frappé*.

Some liqueurs like Sambuca, or drinks like cherry brandy with green Chartreuse floated on top, can be set alight. These create atmosphere and encourage sales in restaurants. Others, like Tia Maria, are often embellished with a layer of cream. For real interest and spectacular eye appeal, there is nothing to beat the *Rainbow Cocktail* or *Pousse Café*. This is a seven drinks creation, built up in equal layers in a liqueur glass. The liqueurs are layered according to their specific gravity, each being poured carefully over the back of a teaspoon. A good combination that works starts off with *grenadine* in the bottom of the glass, followed by: *crème de menthe*, *Parfait Amour*, *Tia Maria*, *blue curaçao*, *green Chartreuse* and *brandy*. However, there are many more combinations you can experiment with.

COCKTAILS AND MIXED DRINKS

Cocktails were first concocted in America and have always been popular there. They came to Britain in the 1920s and had a flourish for about 20 years before losing popular appeal. The increasing affluence of the 1980s brought changes in social and drinking habits. Many older people stopped going out for a drink in the evenings. The young had money and took their place, but many wanted something more glamorous and colourful to drink than ale and stout. So the cocktail, that drink of nostalgia and instant sophistication, re-emerged. Cocktail bars and high profile clubs proliferated. Pubs, bistro/bars and cafés jumped on the bandwagon of this success and introduced the elastic *happy hour* or *cocktail hour* when cocktails and certain nominated drinks could be bought at a reduced or half price.

— *Origin* —

The origin of the word *cocktail* is obscure, but here is one novel notion. In 1779 during the American War of Independence an Irish lady, Betsy Flanagan, kept a tavern much frequented by French and American officers. One day she stole some chicken from her pro-British neighbour which she prepared for the evening meal. She kept aside the feathers from the cock's tail and, as each officer entered, he was offered his favourite mixed drink decorated with a feather. One delighted French officer was so impressed he proposed the toast '*Vive le cocktail*'.

— *The cocktail* —

The base of a cocktail is usually a spirit – gin, whisky, vodka, rum, brandy, tequila or Calvados. The more common flavour contributors are liqueurs, fruit juices, syrups, flavoured fortified wines, eggs, cream, angostura, orange and peach bitters.

There are three basic styles of cocktail: those that require *shaking*, those that require *stirring*, and those that require *layering* or *floating*:

(a) **Shaken** These are cocktails which contain thick or cloudy ingredients such as eggs, cream, syrup and fruit juices.
(b) **Stirred** These contain ingredients which are mostly thin and clear.
(c) **Layered** The ingredients are poured over the back of a spoon into a glass to form layers according to the specific gravity of the ingredients. *Pousse Café* (Rainbow Cocktail) is a good example of such a cocktail.
   Floated One of the ingredients is poured over the back of a spoon onto the top of the cocktail. For example, a layer of Galliano is floated on top in the making of a *Harvey Wallbanger* cocktail.

— *Equipment* —

Professional barstaff will have to hand all the essential equipment necessary to create drinks that will please the eye and excite the palate.

Their panache, flamboyance and skill will add another dimension to the drink and another degree of excellence to the taste.

At home the host or hostess with more limited equipment, less work space and usually less drink choice can also create drinks that have eye appeal and total taste acceptance. The main thing is to keep things simple – experimenting in front of your guests or substituting ingredients because of unavailability are signposts for disaster.

Below is a reasonably selective list of equipment:

- standard cocktail shaker (3 parts)
- Boston shaker (2 parts): especially useful because of its large capacity for cocktails requiring thorough agitation and blending
- bar mixing glass (with pouring lip)
- long mixing spoon
- Hawthorne strainer (with springy edges)
- ice trays, bucket, tongs and crusher
- corkscrew and crown cap opener
- bottle sealer (for Champagne and sparkling wines)
- fruit juice squeezer
- sharp knife and cutting board
- measure/jigger (combining 6 out and 3 out)
- blender/liquidizer
- glasses.

Cocktail decorations

Cocktails can very often look vulgarly overdressed which is off-putting and may also cause difficulties for the drinker. Below is a list of garnishes which can be used in a creative manner to complement the basic flavour of the drink:

- mint and/or borage
- cucumber
- celery stick
- fresh fruit: orange, lemon, lime, pineapple, mango, banana, peach, marachino cherry
- green olive stuffed with pimento
- pearl onions
- cinnamon or nutmeg (for cream-topped cocktails)
- castor sugar or fine salt for frosting glasses.

Frosting glasses

Sometimes a cocktail calls for the glass to be frosted by way of presentation (eg Tequila Sunrise). Glasses can easily be frosted by dipping the rim into a saucer of egg white and then quickly into a saucer of castor sugar. The frosting must be allowed to dry before using the glass. Coloured sugar, made by adding a few drops of food colouring to the castor sugar and blending well, can also be used. If the idea of egg white does not appeal, simply cut an orange or lemon in half and wipe it over the glass rim, then dip the rim into the sugar as before.

— *Tips to ensure success* —

1 Ice should be clean, clear and be the first item to be placed in the shaker or mixing glass.
2 The cocktail shaker should not be filled to more than four-fifths of its capacity (less is better) to allow sufficient room for efficient shaking.
3 Put plenty of ice in the shaker as this helps to chill the drink instantly, as well as giving better balance for a more rhythmic shaking action.
4 When using eggs, check that they are fresh before adding them to the other ingredients. (The use of eggs since the salmonella scare is now a matter of personal choice. The inclusion in this book of drinks containing egg looks forward to a time when the risks associated with eggs, especially raw eggs, have diminished.)
5 Do not put effervescent drinks in a shaker or mixing glass.
6 Handle glasses round the base or stem, never the rim.
7 Never fill glasses to the brim in case of spillage.
8 Whenever possible, serve cocktails in chilled glasses as this helps retain the temperature of the drink.
9 Use the best possible ingredients.
10 Keep to the recipe.
11 When shaking use a short snappy action. Do not prolong the shaking as this merely melts the ice and dilutes the drink. The professionals use the figure of 8, piston, hip and on shoulder styles of shaking.
12 When stirring cocktails, the action should be quick and continuous until the drink is well chilled.

13 Cocktails should be made to order and served immediately when they are at their peak of perfection. Cocktails left standing around will soon separate.
14 Do not over decorate.

Glassware

Glasses should be plain and brilliantly clean. Examples of the most popular glasses used are illustrated and defined below:

Cocktail glasses (smaller): for Pink Lady and White Lady.

Champagne glasses: three styles
   The saucer: for Champagne cocktails and Daisies

The tulip: for Brandy Alexander and Kir Royale

The flûte: for Buck's Fizz and the Grasshopper

Paris goblets: used for Cobblers, Pina Colada, Green Blazer and Whisky Toddy

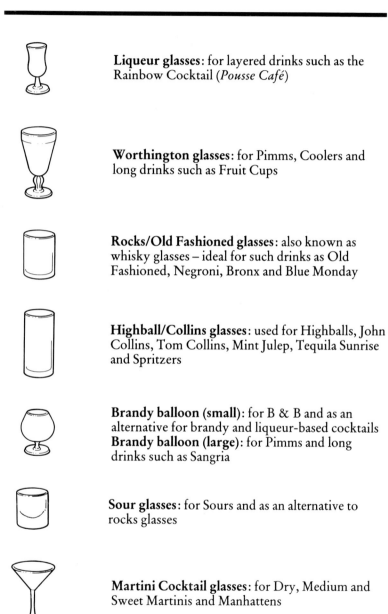

Liqueur glasses: for layered drinks such as the Rainbow Cocktail (*Pousse Café*)

Worthington glasses: for Pimms, Coolers and long drinks such as Fruit Cups

Rocks/Old Fashioned glasses: also known as whisky glasses – ideal for such drinks as Old Fashioned, Negroni, Bronx and Blue Monday

Highball/Collins glasses: used for Highballs, John Collins, Tom Collins, Mint Julep, Tequila Sunrise and Spritzers

Brandy balloon (small): for B & B and as an alternative for brandy and liqueur-based cocktails
Brandy balloon (large): for Pimms and long drinks such as Sangria

Sour glasses: for Sours and as an alternative to rocks glasses

Martini Cocktail glasses: for Dry, Medium and Sweet Martinis and Manhattens

— Cocktail recipes —

Whisky-based cocktails and mixed drinks

Flying Scotsman

2 measures malt whisky
1 measure green Chartreuse
½ white of egg
Shake vigorously with ice
and strain into a cocktail
glass. Decorate with
cherries.

Old Fashioned

1 measure Bourbon
whiskey
1 teaspoon castor sugar
1 teaspoon water
2 dashes Angostura bitters
Stir all the ingredients with
ice in a mixing glass. Strain
and pour into a rocks glass.
Decorate with cherry and
orange.

Mint Julep

2 measures Bourbon
whiskey
6 leaves mint
1 tablespoon fine sugar
Soda water
Put sugar and mint into a
highball glass. Add a little
soda water and thoroughly
mash the mixture until the
sugar is dissolved. Add the
Bourbon and fill the glass
with crushed ice. Stir until
the outside of the glass is
well frosted. Decorate with
mint. Serve with two
straws.

Rusty Nail

1 measure Scotch whisky
1 measure Drambuie
Stir with ice in a mixing
glass. Strain and pour into a
cocktail glass.

Highball

1 measure whisky
dry ginger ale
Put ice into a highball glass.
Add the whisky and ginger
ale to taste. Decorate with
lemon peel.

Manhattan

2 measures rye whiskey
1 measure sweet vermouth
(red)
1 dash Angostura bitters
Stir all together with ice.
Strain into a cocktail glass.
Decorate with a cherry and
lemon peel.

Bobby Burns

1 measure of Scotch whisky
1 measure sweet vermouth
3 dashes Bénédictine
Shake all together with ice.
Decorate with lemon or
orange peel.

Rob Roy

2 measures Scotch whisky
1 measure sweet vermouth
1 dash Angostura bitters
Stir in a mixing glass with
ice. Decorate with a cherry.

Green Blazer

2 measures Irish whiskey
2 teaspoons lemon juice
2 teaspoons clear honey
1 wedge of lime or lemon
studded with cloves
1 very small cinnamon stick
1 teaspoon castor sugar
boiling water
Put all the ingredients in an
8 oz Paris goblet. Add the
boiling water. Stir until
honey and sugar are
dissolved.

Shamrock

½ measure Irish whiskey
½ measure dry vermouth
3 dashes green Chartreuse
3 dashes green crème de
menthe
Stir with ice and strain into
a cocktail glass. Decorate
with a sprig of mint.

Whisky Toddy

Scotch whisky
2 teaspoons castor sugar
boiling water
Put sugar and a little boiling
water into a warmed 8 oz
Paris goblet. Add 2
measures Scotch whisky,
more boiling water, stir and
top up with more Scotch.

Gin-based cocktails and mixed drinks

John Collins

1 measure gin
1 teaspoon sugar
juice of 1 lemon
1 dash Angostura bitters
soda water
Put ingredients except soda water into a highball glass. Stir until sugar is dissolved. Add soda water to taste. Decorate with lemon slice.

Pink Gin

1 measure Plymouth gin
1 dash Angostura bitters
Roll Angostura around a 5 oz Paris goblet until the inside is well coated. Shake out surplus. Add gin and iced water to taste. Some prefer a couple of ice cubes instead of water.

Maiden's Blush

2 measures gin
1 measure Pernod
1 teaspoon grenadine
Shake with ice and strain into a cocktail glass. Decorate with a slice of strawberry.

Horse's Neck

2 measures gin
dry ginger ale
Into a highball glass place ice cubes and gin. Add ginger ale to taste. Decorate with twisted lemon peel.

Gin Sling

1 measure gin
2 teaspoons lemon juice
2 teaspoons castor sugar
1 dash Angostura bitters
Soda water
Put all ingredients, except soda water, into a highball glass with ice. Stir until sugar is dissolved. Add soda water to taste. Decorate with lime and an olive.

Dry Martini

2 measures gin
½ measure dry vermouth
Stir with ice in a mixing glass. Strain into a cocktail glass and decorate with an olive. The amount of vermouth added is a source of constant debate. Some add just a little, others a lot.

Clover Club

2 measures gin
1 measure grenadine
juice of ½ lemon
½ egg white
Shake thoroughly with ice in a Boston shaker. Strain into large cocktail glass. Decorate with lemon peel and a cherry.

Orange Blossom

1 measure gin
1 measure fresh orange juice
Shake with ice and strain into a cocktail glass. Decorate with orange peel.

Pink Lady

2 measures gin
1 teaspoon grenadine
½ white of egg
Shake with ice and strain into a large cocktail glass.

Negroni

1 measure gin
1 measure sweet vermouth
1 measure Campari
Stir all ingredients together in mixing glass with ice. Strain into a rocks glass. Decorate with lemon peel.

Gibson

1 measure gin
1 measure dry vermouth
Shake together with ice. Decorate with an olive and a slice of lime and lemon or with a cocktail onion.

Gimlet

2 measures gin
1 measure lime juice
Shake with ice and strain into a cocktail glass. Decorate with a slice of lime.

Bronx

1 measure gin
½ measure dry vermouth
½ measure sweet vermouth
1 measure fresh orange juice
Shake with ice and strain into a rocks glass. Decorate with a slice of orange.

Tom Collins *is made in the same way as a* John Collins *but uses Old Tom Gin, which is sweeter.*

Singapore Sling

½ measure gin
1 measure cherry brandy
2 teaspoons lemon juice
Shake with ice. Strain into a highball glass and add soda water to taste. Decorate with lime or lemon peel and a cocktail cherry.

Claridge

⅓ measure gin
⅓ measure dry vermouth
⅙ measure Cointreau
⅙ measure apricot brandy
Stir with ice and strain into a cocktail glass. Decorate with orange peel.

White Lady

2 measures gin
1 measure Cointreau
1 measure lemon juice
Shake with ice. Strain into a cocktail glass. Decorate with a cherry on a cocktail stick.

Rum-based cocktails and mixed drinks

Daiquiri

2 measures white rum
1 measure lime juice
½ teaspoon castor sugar
Shake with ice. Strain into a cocktail glass. Decorate with lemon or orange peel.

Blue Hawaiian

½ measure white rum
½ measure blue curaçao
3 tablespoons cream
3 tablespoons coconut cream
3 measures of pineapple juice
Shake with ice and strain into a highball glass. Decorate with pineapple and cherry.

Virgin's Prayer
(serves 2)

2 measures light rum
2 measures dark rum
2 measures Kahlúa (or other coffee flavour liqueur)
2 tablespoons lemon juice
4 tablespoons orange juice
Shake with ice and strain into two highball glasses. Decorate with a grape or strawberry.

Mai Tai

½ measure dark rum
1 measure light rum
½ measure tequila
½ measure Triple Sec
1 measure apricot brandy
1 measure orange juice
2 dashes grenadine
1 dash Angostura bitters
Shake with ice. Put two ice cubes in a Paris goblet and strain in the cocktail. Decorate with pineapple and cherries.

Planters

½ measure golden rum
½ measure fresh lemon juice
1 dash fresh lime juice
Shake with ice and strain into a cocktail glass. Decorate with slices of lime and lemon.

Bacardi

1 measure Bacardi or white rum
juice of 1 lime
½ teaspoon castor sugar
Shake with ice. Strain into a cocktail glass. Decorate with a slice of lime.

Cuba Libre

1 measure white rum
juice of ½ lime
cola
Place rum with lime juice and ice in a highball glass. Add cola to taste. Decorate with slices of lime and lemon.

Pina Colada

2 measures white rum
2 measures pineapple juice
2 teaspoons coconut milk or coconut liqueur
2 dashes Angostura bitters
1 pinch salt
Shake with ice. Pour into a highball glass. Decorate with pineapple, coconut and cherries.

XYZ

1 measure dark rum
½ measure Cointreau
½ measure lemon juice
Shake with ice and strain into a cocktail glass. Decorate with orange and lemon peel.

Vodka-based cocktails and mixed drinks

French Leave

1 measure vodka
1 measure orange juice
1 measure Pernod
Shake with ice. Strain into a
cocktail glass. Decorate
with orange peel.

Moscow Mule

1½ measures vodka
juice of ½ lemon
ginger beer
Fill a highball glass with ice,
add vodka and lime juice.
Top up with ginger beer to
taste. Stir and decorate with
lime.

Black Russian

2 measures vodka
1 measure Kahlúa or other
coffee liqueur
Stir with ice. Strain and
serve in cocktail glass.

Vodkatini

1 measure vodka
1 dash dry sherry
Stir with ice. Strain into a
cocktail glass and decorate
with lemon peel.

Harvey Wallbanger
(Screwdriver with
Galliano)

2 measures vodka
Galliano
fresh orange juice
Fill highball glass with ice.
Add vodka and enough
orange juice to fill the glass
to within 12 mm (½ in)
from the top. Float over the
Galliano on the back of a
spoon. Decorate with an
orange slice.

Screwdriver

1 measure vodka
fresh orange juice
Fill a rocks glass with ice.
Add vodka, top up with
orange juice and stir.
Decorate with a slice of
orange.

Bloody Mary

2 measures vodka
150 ml (6 fl oz) tomato juice
2 dashes Worcester sauce
2 dashes lemon juice
1 dash Tabasco sauce
1 pinch celery salt
A slight sprinkle of cayenne
pepper
Stir together all the
ingredients in a highball
glass with ice. Decorate
with mint and a celery stick.

Blue Monday

2 measures vodka
1 measure blue curaçao
Shake with ice and strain
into a rocks glass.

Tequila-based cocktails and mixed drinks

Acapulco

1 measure tequila
1 measure Tia Maria
1 measure dark rum
125 ml (5 fl oz) coconut
cream
Shake with ice and strain
into a rocks glass.

Tequila Sunrise

2 measures tequila
orange juice
1 teaspoon grenadine
Fill a highball glass with ice
cubes. Add tequila and
enough orange juice to
within 12 mm (½ inch) of
the top. Add the grenadine
and decorate with orange
and a cherry. Add straws.

Margarita

1 measure tequila
1 measure Triple Sec
2 teaspoons lime juice
Shake. Rub rim of cocktail
glass with lemon to
moisten. Dip into a saucer
of fine salt. Strain cocktail
into the prepared glass.

Brandy-based cocktails and mixed drinks

Brandy Alexander

1 measure brandy
1 measure crème de cacao
1 measure cream
Shake with ice and strain
into a Champagne glass.
Sprinkle nutmeg on top.

Brandy Smash

1 measure brandy
1 teaspoon castor sugar
3 sprigs of mint
soda water
In a rocks glass crush mint
with sugar. Add a little soda
water, ice and brandy.
Decorate with a sprig of
mint.

B & B

1 measure brandy
1 measure Bénédictine
Stir in a small brandy
balloon. Ice is optional.

Corpse-Reviver (A)

1 measure brandy
1 measure Fernet Branca
1 measure white crème de
menthe
Shake with ice and strain
into a cocktail glass.

Corpse-Reviver (B)

1 measure brandy
½ measure Calvados
½ measure sweet vermouth
Stir with ice and strain into
a cocktail glass. Add twist
of lemon.

Copacabana

½ measure brandy
1 measure apricot brandy
½ measure Cointreau
2 teaspoons lemon juice
Shake with ice and strain
into a cocktail glass.
Decorate with slices of
orange and lemon.

Between the Sheets

½ measure brandy
½ measure white rum
½ measure Cointreau
1 dash lemon juice
Shake with ice and strain
into a cocktail glass.
Decorate with a strawberry.

Sidecar

½ measure brandy
¼ measure Cointreau
¼ measure lemon juice
Shake with ice and strain
into a cocktail glass.
Decorate with lemon peel.

Liqueur-based cocktails

Grasshopper

½ measure crème de
menthe
½ measure white crème de
cacao
½ measure cream
Shake with ice and strain
into a Champagne glass.

Widow's Kiss

½ measure Bénédictine
½ measure Chartreuse
1 measure Calvados
1 dash Angostura bitters
Shake with ice and strain
into a cocktail glass.
Decorate with a slice of
strawberry and apple.

Honeymoon

1 measure Bénédictine
1 measure Calvados
juice of ½ orange
Shake with ice and strain
into a cocktail glass.
Decorate with a slice of
orange and apple.

Mixed drinks

The main types of alcoholic mixed drinks produced are: *flips, fizzes, noggs, grogs, cobblers, coolers, cups, sours* and *daisies*.

Flips

Brandy Flip

2 measures brandy
1 egg yolk
2 teaspoons
castor sugar

Port Flip

2 measures port
1 egg yolk
2 teaspoons
castor sugar

Sherry Flip

2 measures sherry
1 egg yolk
2 teaspoons
castor sugar

Put the ingredients of the desired flip into Boston shaker with ice. Shake vigorously, strain, and pour into a cocktail glass, sprinkling nutmeg on top. Serve with a straw.

Fizzes

Gin Fizz

2 measures gin
2 teaspoons sugar
juice 1 lemon
soda water

Golden Fizz

2 measures gin
1 teaspoon grenadine
2 teaspoons sugar
juice 1 lemon
1 egg yolk
soda water

Silver Fizz

2 measures gin
2 teaspoons sugar
2 teaspoons fresh cream
juice 1 lemon
soda water

Put the ingredients of the desired fizz into a shaker with ice. Shake vigorously and strain into a Paris goblet. Top-up with soda water and serve with a straw.

Egg noggs

2 measures spirit (whisky, rum, brandy etc)
1 egg
2 teaspoons castor sugar
75 ml (3 fl oz) milk
Put desired spirit, egg and sugar into shaker with ice. Shake vigorously and strain into 200 ml (8 fl oz) Paris goblet. Stir in milk and sprinkle grated nutmeg on top.

Grogs

2 measures spirit (whisky, rum, brandy etc)
2 sugar lumps
2 cloves
1 small stick cinnamon
lemon juice
boiling water
Put the desired spirit, sugar, cloves and cinnamon stick into a Paris goblet with a little lemon juice. Top up with boiling water and stir.

Cobblers

2 measures spirit (whisky, rum, brandy etc)
2 teaspoons sugar
4 dashes curaçao
Half fill a Paris goblet with crushed ice. Add the ingredients, stir and decorate with fruit and a sprig of mint.

Coolers

Rum Cooler

2 measures dark rum
4 dashes grenadine
juice 1 lemon or lime
soda water
Shake all the ingredients together with ice and strain into a highball glass. Add more ice and top up with soda water.

Wine Cooler

1 small glass red or white wine
4 dashes grenadine
soda water
Place the wine and grenadine in a highball glass. Add ice and top up with soda water.

Cups

Claret Cup

1 bottle claret (red Bordeaux wine)
2 tablespoons sugar
juice 1 orange
juice 1 lemon
2 measures orange curaçao
¼ pint drinking water
rind of orange and lemon
Boil the sugar and the lemon and orange rinds in the water. Put these into a container along with the claret, curaçao and fruit juices. Stir and leave in the fridge until ready to serve. Put ice into a glass bowl and pour over the Claret Cup. Decorate with very thin slices of cucumber, apple and orange. Ladle into glasses and decorate with a sprig of mint.

Fruit Cup

3 measures orange squash
3 measures lemon squash
3 measures lime juice
4 teaspoons grenadine
soda water/sparkling mineral water
Half fill a jug with ice and add the fruit squashes, lime juice and grenadine. Top up with soda or sparkling mineral water and stir gently. Decorate with slices of fruit, cherries and a sprig of mint, borage or cucumber rind.

Pimms Cup

large measure Pimms No 1
lemonade/Seven-Up/tonic water
Pour the Pimms into a large bowl-shaped glass such as a brandy balloon or a ½ pint tankard or Worthington glass. Add ice and top up with lemonade or an alternative. Decorate with a cherry and slice of apple, orange, lemon, lime and a twist of cucumber peel. A stirrer and two straws are optional.

Hock Cup

½ bottle Rhine wine
2 measures medium sherry
1 measure curaçao
soda water
Pour the wine, sherry and curaçao into a jug. Add ice, top up with soda water and decorate with a slice of lemon and cucumber rind.

Cider Cup

2 bottles cider
2 measures brandy
1 measure Maraschino
soda water
Pour the cider, brandy and Maraschino into a glass jug. Add ice and slices of assorted fruit. Top up with soda water. Stir slightly and decorate with apple slices and cucumber rind.

Sours

Whisky Sour

2 measures whisky
1 measure lemon juice
1 dash gomme syrup/1 teaspoon castor sugar
1 dash egg white
Shake all the ingredients with the ice and strain into a rocks glass. Decorate with a slice of lemon.

Frisco Sour

2 measures Bourbon whiskey
1 measure Bénédictine
1 measure fresh lime juice
1 measure fresh lemon juice
Shake with ice and strain into a rocks glass. Decorate with a slice of lemon and lime.

Daisies

2 measures spirit or liqueur (rum, whisky, brandy, cherry brandy etc)
1 measure grenadine
juice of ½ lemon
Put all the ingredients into a shaker, together with some ice. Shake and strain into a Champagne saucer. Top up with a little soda water and decorate with cherries.

Wine-based cocktails and mixed drinks

Black Velvet

½ chilled Champagne
½ Guinness
Pour the Champagne and Guinness simultaneously into a chilled silver tankard, taking great care to avoid frothing over.

Kir

chilled dry white Burgundy, eg Chablis or Aligoté
1 teaspoon crème de cassis
Put the crème de cassis in a goblet and pour over the chilled white wine.

Kir Royale

As above, but substitute chilled Champagne for white wine.

Champagne Cocktail

Champagne
1 sugar lump
Angostura bitters
1 teaspoon Cognac/orange curaçao
Dampen the sugar lump with Angostura and place in a Champagne glass (the saucer style if you prefer). Pour over well chilled Champagne and float the Cognac or orange curaçao over the back of a teaspoon. Decorate with a slice of orange and a cocktail cherry or, more simply, with a strip of orange peel.

Buck's Fizz

1 measure chilled, fresh orange juice
1 dash grenadine
chilled Champagne
Stir the orange juice and grenadine together in a wine glass. Top up with Champagne (it should be Bollinger, but other brands can be substituted). Decorate with a slice of orange or a strip of orange peel.

Sangria
(serves 12)

1 bottle reasonable quality Spanish red wine
3 measures brandy
125 ml (¼ pint) orange juice
500 ml (1 pint) lemonade
Pour all the ingredients, together with some ice, into a glass bowl or other glass container. Stir until cold. Decorate with thin slices of orange, lemon and lime.

Spritzer

½ glass white wine
soda water or sparkling mineral water
Put the wine and ice into a highball glass. Top up with soda water or mineral water and stir gently.

Mulled wine and winter warmers

Mull of Mayo
(serves 20)

2 bottles Burgundy/Rhône red wine
¼ bottle dark rum
½ bottle Dubonnet
½ bottle drinking water
1 orange liberally studded with cloves
2 cinnamon sticks
25 g (1 oz) sultanas
2 lemon halves
5 g (¼ oz) mixed spice
1 400 g (1 lb) jar clear honey
Heat the orange in the oven for 10 minutes to bring out the full flavours. Tie the mixed spices securely in a muslin bag to prevent cloudiness, so that only the flavour will be released. Then place all the ingredients, except the rum, in a large pot – do not use all the honey so the flavour can be adjusted later. Place the pot on a low heat and stir occasionally. Bring the mixture to boiling point, but do not allow to boil. Add the rum, stir, taste and add more honey if necessary. Return to boiling point and then serve into goblets using a ladle with a lip. Sprinkle a little nutmeg over each drink.
　Serve immediately whilst it is fresh and hot. (Tepid mulled wine is insipid.)

Dr Johnson's Choice
(serves 12)

1 bottle claret (red
Bordeaux wine)
1 wineglassful orange
curaçao
1 wineglassful brandy
sliced orange
12 lumps sugar
6 cloves
1 pint boiling water
Heat the wine with the
orange slices, cloves and
sugar until nearly boiling.
Add the boiling water,
curaçao and brandy. Pour
into glasses and sprinkle
grated nutmeg over the top
of each drink.

The Bishop
(serves 12)

1½ bottles ruby port
50 g (2 oz) lump sugar
2 oranges
2 cinnamon sticks
cloves
1 pint water
Prick one of the oranges all
over with cloves. Place this
in a medium oven for about
half an hour.

Pour the port into a
saucepan and bring to
boiling point, but do not
allow to boil. Meanwhile,
boil the water with the
cinnamon sticks and the
baked orange. Rub the
sugar lumps against the skin
of half of the second orange
and place in a serving bowl
with the juice of the orange.
Combine the heated port
and boiling water and pour
into the serving bowl.
Allow the cinnamon sticks
to remain in the bowl,
together with the orange
'hedgehog' for decoration.

In the absence of a suitable serving bowl, the mulled wines and winter warmers can be stored in bottles placed in a basin of hot water. When serving, a clean napkin should be wrapped around each bottle.

5 Non-alcoholic drinks

MINERAL WATERS

There has been an enormous increase in the consumption of mineral water in recent years. Whether this is because people no longer trust the tap (due to the well publicized deterioration in the quality of our water) or whether they are concerned about their diet and general state of health, or simply because they enjoy the taste of a favourite brand is hard to tell. However, mineral waters are now very 'vogue' to drink.

The main sales outlets for mineral water are supermarkets, grocers, off-licenses and, more recently, restaurants. The latter are now, not before time, including them on their wine lists, albeit in some cases at astronomical mark-ups.

In Britain sparkling waters account for about 65% of mineral water sales. To ride on the back of this success many traditional still water producers, like Malvern, have now launched a sparkling variety on the market. Many firms, at present, produce both the still and sparkling varieties. The sparkle is either natural or induced – when natural the producer will be happy to make that claim on the label.

— Brand names —

The most popular brand names associated with mineral water, together with their country of origin are listed opposite.

— Service of mineral waters —

In resturants, the sommelier will usually suggest a mineral water after the wine order has been taken. Should the customer not require a mineral water, then a jug of iced water would be offered. This is part of good service which customers appreciate.

Country	Brand name	Country	Brand name
France	Badoit	Ireland	Ballygowan
	Contrex		Glenpatrick
	Evian		
	Perrier	Scotland	Highland Spring
	Vichy Saint Yorre	Wales	Breckon Saint David
	Vittel	Italy	San Pellegrino
	Volvic		Ferrarelle
England	Ashbourne		Crodo Lisiel
	Abbey Well	West Germany	Apollinaris
	Ashe Park		Überkinger
	Malvern		Petrusquelle
	Aqua Pura	Belgium	Spa
		Switzerland	Aqui
		Sweden	Ramlosa

Mineral waters should be nicely chilled, taken straight from the fridge and opened only in front of the customer. The water should be poured into a sufficiently large stemmed glass (an 8 oz Paris goblet is ideal). No ice, lemon slice or garnish of any kind is required.

SYRUPS

A syrup comprises a solution of sugar and fruit juice. They are used mostly in cocktail creations and general bar work, although they are sometimes of use in the kitchen. The most widely used syrups are:

sirop de cassis (blackcurrant)
sirop de cerises (cherry)
sirop de framboises (raspberry)
sirop de gomme (gum/sugar)

sirop de grenadine (pomegranate)
sirop de groseilles (gooseberry)
sirop de oranges (orange)
sirop de orgeat (almond)

MINERALS AND MIXERS, JUICES AND SQUASHES

A huge industry has been built around these drinks, especially since the introduction of low calorie varieties. Coca-Cola and Pepsi Cola, Fanta, 7-Up, Schweppes and Lilt are well-known brands. Where possible and practical these drinks should be opened in front of the customers and garnished according to flavour. They are either drunk on their own or are used as mixers with spirits.

- Juices should be well shaken before being opened. Orange juice combined with bitter lemon makes a refreshing drink called *St Clements*.
- A combination of lemonade, cola and ice, called *Spezi*, is a delicious thirst quencher.
- Squashes combined with ice and water are used for long cooling drinks. They are also the principal ingredients with grenadine for colourful fruit cups.
- The gin-based *Pimms No 1* is usually topped up with lemonade. As an alternative to lemonade, tonic water or a personality mineral water such as Ballygowan may be used.
- Tomato juice should be poured into a glass at the table and Worcester sauce offered with a teaspoon on a side plate for use by the customer if required.

6 Cigars

Although cigars are nowadays made in a host of countries like the **USA**, **Puerto Rico**, the **Philippines**, **Japan**, the **Dominican Republic**, and the **East Indies**, the true home of the cigar is **Cuba** and, to a lesser extent, **Jamaica**.

CIGAR COMPONENTS

There are three parts to a good cigar: the *filler*, the *binder* and the *wrapper*.

Filler

The filler comprises the bulk of the cigar and is made of a blend of leaves to form the inner core. This blend gives the cigar most of its flavour, so quality here is important.

Binder

The binder, the inner covering of the cigar is made of a single quality leaf which binds the filler. These together form what is known as the *bunch*.

Wrapper

The wrapper is an exceptionally fine single leaf which must have, as requirements, *elasticity*, *strength* and a *fine appearance*. The wrapper leaf is classified according to colour:

Claro	(CCC) light coloured
Colorado Claro	(CC) medium
Colorado	(C) dark
Colorado Maduro	(CM) very dark
Maduro	(M) exceptionally dark

STORAGE

Cigars should be stored at a temperature of between 15 °C and 18 °C, (16.5 °C is best), with a relative humidity of between 53% and 57%. Cedar wood boxes or cedar lined containers are ideal, as cedar, being porous, allows the cigar to breathe and the aroma of cedar blends well with that of a cigar. Using a humidor is also a good way of keeping cigars in condition. Sometimes when cigars are badly stored a greyish mildew or grey specks may appear on the wrapper. These can be wiped away quite easily with a soft brush. They are not harmful and neither are the yellow and green spots which you sometimes see. The yellow spots occur through the sun drying rain drops on the tobacco leaves as they grow and the green indicates an over abundance of oil. Both demonstrate the authenticity and naturalness of the tobacco leaf.

SERVICE

Cigar boxes should be opened carefully with a blunt instrument. In a box of 25, 13 cigars should be on the top and 12 on the bottom layer. To extract a cigar, press the rounded head and the cigar will tilt upwards for easy extraction. The band or identification tag is best removed immediately as it can damage the outer leaf if moved up and down.

When cigars are not pre-cut a V-shaped cigar cutter is required to cut the end (in a V), thereby facilitating maximum free draught and ease of smoking. Do *not* make a small hole with a match or cocktail stick as this will leave a moist tar concentrate which imparts a very bitter flavour as you approach the end of your smoke.

Light cigars with the broad flame of a match, with a cedar wood spile or with a gas lighter, rotating the cigar to effect even burning. Should the cigar need re-lighting, first remove all excess ash and then blow through it onto the flame.

When a cigar is being carried for smoking later it is best to use a tube or a special leather case with separate tubings. Some people carry cigars along their inner leg, inside their sock.

CIGAR SIZES

Corona	14.5 cm (5½ in) with a round top
Petite Corona (*Corona Chica*)	13 cm (5 in) with a round top
Très Petite Corona	as *corona* but 11.5 cm (4.5 in) in length
Half Corona	as *corona* but 9.5 cm (3.75 in) in length
Lonsdale	as *corona* but about 16.5 cm (6.5 in) in length
Ideales	torpedo shaped, about 16.5 cm (6.5 in) in length
Londres	straight cigar, 12 cm (4.5 in) in length
Panatella	long and thinnish and open at both ends, 12.5 cm (5 in) in length
Stumpen (*Cheroot*)	stubbier than a *panatella* but slightly tapered, open at both ends
Whiff	Usually small and open at both ends, about 8.9 cm (3.5 in) in length

7   Glossary of wine and drink terms

Abboccato	Italian for 'soft caressing' and associated with slightly sweet wines
Abfüllung	bottling
Acerbe	immature acid wine
Acid	imparts lasting qualities, adds bouquet and flavour; too much acid makes wine sharp or sour; too little makes wine flat
Adegas	Portuguese warehouse for storing wine
Age	refers to the maturing of wine
Agrafe	iron clasp which holds down the first cork on a Champagne bottle
Aigre	sour wine (*vinaigre* – vinegar)
Ainé	senior partner in a wine firm
Albariza	chalky soil found in the best sherry country of Spain
Alcohol (ethyl)	C_2H_5OH, obtained by the action of yeast on sugar during fermentation. Its strength can be further increased by distillation
Aldehyde	half way stage between a wine and an acid
Alembic	a still
Aligoté	big yielding white grapevine in Burgundy
Amer	bitter
Amer picon	bitter-tasting French aperitif
Ampelographer	expert on grape vines
Añada	vintage wine in Spain
Anbaugebiet	German for wine region
AOC	(Appellation d'Origine Contrôlée) – highest quality control for French wines

Appellation contrôlée	French law which guarantees the origin of the wine named on the label (also called *Appellation d'Origine Contrôlée*)
Âpre	harsh
Aqua vitae	'water of life' or 'spirit'; similar to eau de vie and uisge beatha
Arenas	sandy soil in southern Spain
Argol	tartaric deposit thrown as wine matures in cask
Aromatized wine	a flavoured wine such as vermouth, usually fortified
Arrope	boiling down of *must* to sweeten and add colour to sherries
Asciato	Italian for dry wine
Atmosphere	applied especially to sparkling wine. It is a measure of atmospheric pressure. One atmosphere is equal to 15 pounds per square inch. A Champagne bottle usually has an internal pressure of six atmospheres
Auslese	specially selected grape bunches for making some German wines
Balderdash	mixture of drinks which are generally unrelated, eg wine and milk. (It is also an Irish term to describe illogical conversation)
Barrel (wine)	holds 26.25 gallons (119.5 litres), but it can vary
Barrel (beer)	holds 36 gallons (163.6 litres)
Barrique	barrel usually holding about 225 litres. Barrique wines are oak aged wines
Barros	clay soil in southern Spain
Baumé	a measure indicating the sugar content in *must*
Beer	fermentation of barley, malt and hops
Beerenauslese	individually selected ripe grapes for making some German wines
Beeswing	floating sediment that has not settled with the crust in vintage port
Bentonite	a special, fine clay which is often used as a fining agent to clear wine
Bereich	a German wine district

Bianco	white
Blackstrap	used long ago to describe a rough variety of port. Elderberry colouring was usually added to disguise faults
Black Velvet	a mixture of chilled Champagne and Guinness. Sometimes called 'Bismark'
Bin	slot for holding a bottle of wine in a cellar
Blanc de blancs	white wine from white grapes
Blanc de noirs	white wine from black grapes
Blanco	white
Blush wine	pale pinky blue wine from black grapes
Bock	beer tankard made of glass which holds about ½ pint (285 cm^3). Used in France for drinking draught beer
Bocksbeutel (boxbeutel)	attractive flagon-shaped bottle for holding the Steinwein of Franconia. Now used extensively in Portugal
Bodega	Spanish cellar, warehouse or bar
Body	description of strength and fullness in wine
Bois	wood – *Gout de Bois* – woody taste
Bonded	wines and spirits are bonded and kept in a warehouse under government supervision until the Customs and Excise duties are paid by the purchaser
Bota	Spanish for butt or cask. Holds 108 gallons (490 litres)
Bottle	standard size holds 75 cl
Bottle sickness	sometimes happens to newly bottled wine. Disappears after some months
Bottoms	lees or dregs left after racking or decanting. In a 36 gallon (163.6 litre) barrel of beer, the natural sediment – yeast and hop débris – amounts to about 1 gallon (4.5 litres)
Bouche	mouth
Bouchonné	corked
Boude	bung
Bouquet	aroma, smell or nose of wine
Botrytis cinerea	*pourriture noble* – noble rot
Brandewijn (branntwein)	burnt wine – brandy

Breed	name used to describe fine quality wine
Brut or *nature*	driest Champagne – generally no sweetness added
Butt	cask holding 108 gallons (490 litres)
BOB	(Buyers Own Brand) – wine or spirit made to be sold under the name of restaurant or supermarket
Cabinett (kabinett)	to describe wine good enough for the German vineyard proprietor's own cellar
Capataz	head cellarman of a Spanish cellar
Capiteux	heady wine, high in alcohol
Cask	wood container for wines, beers, spirits
Cava/cave	cellar
Caviste	cellar worker
Cellar	below ground storage area
Cep	vine stock
Cépage	vine variety
Chai	above ground storage area
Chambrer	to bring wines (usually red) to room temperature
Chaptalization	addition of sugar to grape must to secure higher alcoholic content. Amount is strictly controlled by law
Charnu	a full bodied wine
Château	castle – also means a wine from a particular vineyard
Château bottled	*Mise en bouteille au château* – signifies that the wine has been bottled at the château, which in itself is a guarantee of quality
Chlorosis	a vine disease caused by an imbalance of minerals
Claret	English name for the red wines of Bordeaux. Comes from the French *clairet* meaning clear, bright, light
climat	vineyard or single plot within a large vineyard
Clos	walled vineyard, especially in Burgundy
Cobblers	American drink for warm weather. Made from wine or spirits, fresh fruit and ice shavings

Collage	clearing wine of its sediment – fining
Copita	Spanish sherry glass
Cork	usually made from the bark of Spanish or Portuguese oak (*Quercus suber*)
Corky	when a wine has been diseased by a faulty cork
Cortado	a sherry, between an amontillado and oloroso in style
Corsé	full-bodied
Côtes	hills where some vineyards are located
Coulure	too much rain and soil deficiency is the principal cause. The berries on the vine will not develop and stay stunted
Coupage	vatting or blending of wine
Crémant	creaming, sparkling, effervescent
Criadera	nursery for young sherries
Cru	growth – also wines of a similar standard
Crust	deposit which has gathered especially in bottles of vintage port
Cuit (cotto)	'cooked' wine
Cups	long mixed drinks usually made in glass jugs
Cuve	vat
Cuve close	bulk method for making sparkling wine in a sealed tank. Also known as the Charmat method after its inventor Eugène Charmat (1907)
Cuvée	blend of wines
Dame-Jeanne (demijohn)	a wide-waisted covered glass wicker jar which holds 13.5–45.5 litres. Used for storing Madeira
Decanter	glass container of different shapes and often highly ornamented
Decanting	the transference of the liquid from bottle to decanter
Dégorgement	release of sediment especially in Champagne bottles
Demi-sec	half dry, fairly sweet
Depôt	sediment
Diastase	the enzyme complex which converts the starch in malted barley into invert sugar

Distillation	the application of heat which extracts alcohol from fermented liquids
DO	*Denominación de Origen* – Spanish equivalent to *AOC* of France
DOC	*Denominazione di Origine Controllata* – Italian equivalent to *DO* and *AOC*
DOCG	*Denominazione di Origine Controllata e Garantita* – the highest and newest classification of Italian wines
Doble pasta	associated with some Alicante red wines. Double the normal amount of grape skins is left with the must during fermentation
Domaine	privately owned vineyard in Burgundy
Dosage	addition of sugar dissolved in wine to produce varying degrees of sweetness in sparkling wines
Doux	Sweet, also *dulce*
Dur	hard
Eau de vie	similar to aqua vitae and uisage beatha. Refers to spirits and means 'water of life'
Echt	genuine or right
Edel	noble
Edelfäule	overripe almost rotten grapes – 'noble rot'
Edelgewächs	best vintages
Égrappoir	machine used to de-stalk grapes before pressing
Eiswein	ripe grapes which have become iced or frozen while on the vine. Seldom occurs but makes delicious wine
Einzellage	single vineyard
Eleveur (élevage)	the 'bringer up' of wine. These terms refer to the *négociant*, the person who buys the young wine from the farmers and looks after it until it is ready to be sold, usually under the négociant's name
Elixirs	liqueurs made from the finest distillates
Esters	combination of acids and alochol which gives wine its bouquet
Estufa	hot chambers where young Madeiras undergo heating or baking

Ethyl alcohol	the main alcohol in wine
Extra sec	less dry than *brut*
Faible	thin wine
Fass no (füder no)	cask number
Faul	German for 'mouldy', 'rotten' or 'foul'
Fein	fine; *feinste* – finest
Feints	the last part of spirit to emerge from a pot still. The first are 'foreshots'. The in-between is called the 'heart'
Fermentation	action of yeast on sugar which converts the latter into roughly equal parts of alcohol and carbon dioxide
Fiasco	wicker-wrapped bottle used for Chianti and some other Italian wines
Filtering	making wine bright
Fine	brandy of no great distinction
Fine Champagne	finest Cognac brandy
Fine maison	brandy of the house or restaurant
Fining	clearing wine in cask or tank. Isinglass, whites of egg, gelatine and some clays are used for the purpose
Fino	driest sherry
Flaschenschild	label
Flor	'flower' – yeast scum growth on some sherries maturing in cask in Spain
Fort	strong
Fortified	strength added by addition of grape spirit
Foudre	huge casks for storing wine
Franc	clean-tasting
Frappé	iced
Frisch	fresh
Fruité	fruity
Füder	German cask
Fumet	definite bouquet
Fürst	prince
Fusel oil	alcohol (not ethyl) found in spirits
Gallon	4 quarts, 8 pints, 4.54 litres, 1.2 American gallons
Garrafa	decanter

Gay-Lussac (GL)	French scale for measuring alcoholic strengths. Now known as *OIML*
Généreux	generous wine – in alcohol. A fortified sweet wine
Geropiga	natural port or special liquid made to sweeten other Portuguese wines
Gewächs	vineyard of, or growth of
Glühwein	mulled wine, spiced and sweetened
Goldbeerenauslese	picked individual bunches of fully ripened grapes
Goût	taste
Goût Américain	fairly sweet
Goût Anglais	dry
Goût de bouchon	corky tasting wine
Goût d'évent	flat
Goût Français	sweet
Goût de moisi	musty
Goût de paille	straw
Goût de piqué	going towards vinegar
Goût de taille	uncouth, made from final pressing of grapes
Goût de terroir	earthy
Grosslage	a wine area that is part of a *bereich*
Growth	can mean a 'vineyard'
Heads	first spirit to emerge during distillation
Hectare	2.47 acres
Hectolitre	liquid measure equalling 22 gallons (100 litres)
Hock	Rhine wines – abbreviation of *Hochheim*
Hogshead	cask of different sizes ranging from 48–60 gallons (221 to 272.6 litres)
Hors d'age	should it appear on a brandy bottle it means 'beyond recorded age'
Hybrid	a cross between two grape species
Hydrometer	an instrument to record the density of alcohol in a wine or spirit
Impériale	large bottle – holds between six and nine bottles. Claret is sometimes matured in such bottles
Isinglass	fish gelatine often made from the bladder of the sturgeon, used for fining

Jahrgang	vintage, year
Jarra	jar of varying sizes
Jeroboam	4 bottles; a Champagne bottle
Jeropiga	boiled-down grape juice, used in Portugal to sweeten wines
Jigger	1½ oz measure (4.26 cc)
Kabinett	genuine unsugared wine. First grade of *QmP* wines
Keg	cask
Keller	German for cellar
Kellerabfüllung (kellerabzug)	estate bottled
Lagar	trough for pressing grapes in Spain and Portugal
Lage	site
Lagered	storing of beer during which time it is fined and carbonated
Landwein	an honest country wine, of no great distinction
Lees	sediment, left in cask or vat during the fermentation period
Léger	light
Levante	hot winds in the sherry district of Spain
Limousin	special oak for maturing Cognac
Liqueur d'expédition	the sugar and old wine added to Champagne after the disgorging
Liquoreux	heavy sweet wine
Lodge	where port and Madeiras are stored
Maceration	when grape skins are left with the must during fermentation
Maderization	caused by too much oxygen during fermentation or maturation. It turns white wines brown in colour and flat to taste
Maestro vino	master wine used for adding colour to Málaga wine
Malt	grain which has germinated
Marc	brandy from the 3rd or 4th pressing of the grapes

Marque	a brand
Mash	mixture of ground malted barley or other cereal and water
Mesa	table wine
Méthode Champenoise	second fermentation in bottle (Champagne method)
Méthode cuve close	making sparkling wine by tank method
Microclimate	a special, usually favourable, climate particular to a small area or even one vineyard. It could depend on shelter, exposure or general location – mountains, hillsides, water etc
Mildew	a vine disease which occurs mostly in damp weather
Millésime	vintage date in France
Mistelle	*must* which has had its fermentation stopped by the addition of brandy at a very early stage
Morgen	a German acre of land
Mouillé	watered
Moût	unfermented grape juice
Mousse	effervescence in a sparkling wine
Mousseux	sparkling
Mou vin	lifeless wine
Muffa nobile	noble rot
Mûr	ripe
Must (most, mosto)	unfermented grape juice
Mut	balanced
Mutage	adding pure alcohol to *must* during fermentation to stop fermentation – as in the case of *vin doux naturel*
Natur (naturwein)	A wine with no sugar added, a natural wine
Négociant (eleveur)	wine handler or firm who buys young wine from the producers and matures and bottles it for sale under his own label
Nero	black or deep red wine
Nerveux	a strong, full-bodied wine
Nicolauswein	wine made from grapes gathered on St Nicholas' Day – 6 December

Nip	very small bottles of spirits or Champagne (split)
Nu	bare – cost of wine without its overheads (cask, bottles, etc)
Nube	cloudiness
Obscuration	amount of false reading on a hydrometer caused by impurities (mainly sugar) in the alcohol
Œchsle	is a standard which determines the specific gravity of a given *must*. It will show the number of grams by which one volume of *must* is heavier than an equal volume of water, the sugar content being in the region of 25% of this measurement. So, 100 litres of *must* with a reading of 100° *Œchsle* will contain some 25 kilogrammes of sugar
Oeil de perdrix	tawny pink wine, colour of 'partridge eye'
Oenology	study of wine from a scientific point of view
Offener wein	wine sold by the glass
Oidium	fungus disease on vines
OIML	European scale for measuring alcoholic strength (based on the French *Gay-Lussac* scale)
Oloroso	heavy golden sherries which are dry in their natural state
Ordinaire	wine for everyday use, usually cheap
Originalabfüllung	German equivalent to estate bottled or *mise au château*
Organic wines	a generic name for wines produced from grapes that have been cultivated naturally – without pesticides, herbicides or chemical fertilizers
Palma	classification of young fino sherries
Parfume	fragrance or perfume in wine
Passe-tous-grains	Burgundy blend of ⅓ *Pinot Noir* to ⅔ *Gamay* grapes
Pasteur, Louis	French scientist renowned for his work on pasteurization and fermentation
Pastis	French name for aniseed-flavoured aperitifs

Patent still	a continuous still patented by Irishman Aenaes Coffey in 1832
Paxarette	a sweetening wine made from the *PX* grapes
Pedro Ximénez (PX)	grapes in the sherry country of Spain
Pelure d'oignon	colour of onion skin. Red wines which gradually take on this colour due to their great age
Perlwein	effervescent wine
Pétillant	semi-sparkling or creaming wine
Petit	small
Phylloxera vastatrix	vine louse from America which devastated European vineyards in the 1870s
Pièce	Term used chiefly in Burgundy. See *Hogshead*
Pint	liquid measure of 20 oz (568.2 cm^3)
Pipe	usually applies to a cask of port holding about 115 gallons (522.5 litres)
Piqué	pricked
Piquant	sharp-tasting wine
Pisador	person who treads grapes in sherry region
Plastering	addition of gypsum (calcium sulphate) to grapes when they are being pressed
Portes greffes	American vine stocks resistant to *Phylloxera* on which European stock is now grafted as a remedy against the pest
Pourriture noble	noble rot – grapes left on vine until they become like raisins, similar to *Edelfäule* in Germany
Précoce	a wine which has come forward too early – a precocious wine
Pressoir	maturing wine press
Proof	method of measuring alcoholic strength of a liquid
Punt	the dip in the bottom of a bottle. Adds extra strength and when wines throw sediment, it helps to retain this at the bottom of the bottle
Pupître	a wooden frame for holding Champagne bottles during the *remuage* process
Putton	Hungarian bucket for gathering grapes in the Tokay district. Holds about 13.5 kilos

QbA	*Qualitätswein bestimmter Anbaugebiet* – a quality wine from a specific region
QmP	*Qualitätswein mit Prädikat* – a quality wine with certain distinctions
Quartaut	barrel holding about 56 litres
Queue	Burgundy cask holding 2 hogsheads
Quinta	Portuguese estate or vineyard
Race	breed
Racking	changing wine from one cask to another, leaving the sediment behind in the old
Raki	Balkan countries' name for a spirit flavoured with liquorice and aniseed
Ratafia	style of liqueur made from must and marc
Raya	system for classifying sherry
Rectify	to change a natural spirit in some way, either by redistilling or adding colour and flavourings
Refractometer	an instrument or optical device used to gauge the sugar content in grapes while still on the vine. It indicates if the grapes are ready for harvesting
Refreshing	adding younger wine to older
Rein	pure
Remuage	the guiding of sediment down to the neck of the bottle. Associated with the production of Champagne
Robe	French for the colour of wine
Roundeur (rund)	round
Rosé	pink wine
Rotwein	red wine
Ruby	young, deep-red port
Saccharometer	an instrument for measuring sugar content in *must* or wine
Sacramental wine	altar wine for the Eucharist. May be red or white and must be the natural grape wine
Sack	anglification of Spanish Seco, a dry fortified wine
Sancocho	boiling must down to ⅓ of its volume. This is used to sweeten and colour sherries

Scantling	wooden beams to support casks in cellars
Schloss	German for castle
Schlossabzug	estate bottled
Schnapps	German, Dutch name for spirit
Schneewein or *Eiswein*	snow wine
Schorle-morle	mixture of wine and aerated mineral water
Sec (secco)	dry
Sediment	deposit in wines as they age
Sekt	German sparkling wine
Sève	flavour and body of wine which augurs for it lasting well
SO₂	formula for sulphur dioxide
Solera	system for blending and maturing sherries to give a consistent standard
Soutirage	name in Champagne for racking
Soyeux	smooth, silky
Spätlese	wine made from late gathered grapes
Spitzengewächs	best German growths
Spitzenweins	best wines
Spritzig	slightly effervescent wine
Spumante	Italian for sparkling
Staatsweingut	vineyard owned by the state
Still	an apparatus either of the 'pot' or 'patent' variety where fermented wash or wine is distilled into a spirit
Still wine	non-sparkling wine
Stück cask	measure in Rhineland holding 1,200 litres
Sur lie	wine completely matured with its lees before bottling
Süssreserve	unfermented grape juice used to freshen and sweet-flavour some German wines
Sykes hydrometer	the hydrometer for measuring alcoholic strength, named after the inventor
Tafelwein	table wine (German) also called *Vin de Table* (Fr), *Vino da Tavola* (It) and *Vino de Mesa* (Sp)
Tannin	astringent acid in wine found in stalks, pips and skins of grapes

Tappit-hen	Scotch bottle which holds three imperial quarts which is equivalent to 4½ reputed quarts
Tastevin	dimpled silver instrument or cup used to taste wine. Qualified sommeliers (wine waiters) sometimes wear them on a neck chain in restaurants
Tent	Spanish for sweet wine
Tête de cuvée	best growth, best wines from any vineyard
Tinaja	large earthenware jar
Tinturier	grapes for colouring
Tirage	bottling
Tonelero	cooper
Traube	grape
Traubensaft	grape juice, *must*
Trocken	dry
Uisage beatha	similar to aqua vitae and eau de vie
Ullage	wine lost through evaporation or leakage. Describes also the air space in cask above fluid
Usé	wine past its best
Vat	large casks for blending and maturing wines and spirits
VDN	*Vin Doux Naturel* – sweet wine that has been muted with brandy when the alcohol level reached between 5% and 8% by volume
VDQS	*Vin Délimité de Qualité Supérieure* – wine of superior quality produced in a demarcated region
Velouté	velvety
Vendange	vintage, harvesting of grapes
Vendangeur	vintage worker
Venencia	silver cup attached to a long whale-bone handle for sampling sherries from cask
Verbesserung	sugaring of wine in a poor year to increase alcoholic content
Vert	green, young wine
Viejo	old
Vigne	vine

Vigneron	vineyard worker
Vignoble (vinedo)	vineyard
vin (vinho, vino, wein)	wine
Vin de garde	wine for laying down
Vin de goutte	poor-quality wine from last pressing of grapes
Vin de messe	altar wine
Vin de paille	straw wine. Grapes have been allowed to dry out on straw
Vin de pays	small local wine
Vin doux	sweet wine
Vine	plant on which grapes grow
Vintage	gathering of grapes
Vintage wine	wine of one year
Viscosity	sometimes when a wine is swirled or rolled around a glass, tears or legs form near the top of the glass and run back down into the wine. This indicates a high level of sugar, alcohol or other matter – perhaps a combination of all three
Viticulture	culture of the vine
Vitis	vine genus
Wachstum	growth
Wash	fermented liquor before it goes for distillation
Wassail	from the Anglo-Saxon *weshal* meaning 'be of good health'
Weeping	leaking cork or cask
Weingut	vineyard
Weisswein	white wine
Wine	fermented juice of freshly gathered grapes
Woody	wine with the smell of cork
Würzig	spicy
Yeast	uni-cellular fungi found on skins of grapes which during fermentation act on the natural grape sugar to convert it into alcohol and carbon dioxide

Yeso	calcium sulphate (gypsum) which is sprinkled on grapes during pressing to slow down fermentation and add acidity
Zapatos de pisar	special boots, traditionally used for treading grapes in the Sherry district

— *Recommended Reading* —

World Wine Encyclopedia, Tom Stevenson, London 1989
Liqueurs and Spirits, Henry McNulty, London 1985
The Wine Buyer's Guide, Robert Parker, London 1987
Drinks and Drinking, John Doxat, London 1971
Vendange, Andrew Durkan, London 1971
Cocktails & Party Drinks, Tiger Books Int., London 1989
Which? Wine Guide, Edited by Roger Voss, London 1990
Food and Beverage Service, Lillicrap & Cousins, London 1990